Elementary
Crystallography

Elementary
Crystallography

D. VELMURUGAN

Professor
CAS in Crystallography and Biophysics
and
Dean (Research)
University of Madras
Chennai

MJP PUBLISHERS
Chennai 600 005

ISBN 10: 81-8094-033-0
ISBN 13: 978-81-8094-033-0

Cataloguing-in-Publication   Data
Velmurugan, D (1960 –   ).
    Elementary Crystallography / by D. Velmurugan.
Chennai : MJP Publishers, 2008
    xviii, 306 p. ; 21 cm.
    Includes Glossary, References and Index.
    ISBN 81-8094-033-0 (pbk.)
    1. Crystallography, I. Title.
    548 VEL MJP 036

ISBN 81-8094-033-0   **MJP PUBLISHERS**
© Publishers, 2008   47, Nallathambi Street
All rights reserved   Triplicane
Printed and bound in India   Chennai 600 005

Publisher : J.C. Pillai
Managing Editor : C. Sajeesh Kumar
Project Editor : P. Parvath Radha
Acquisitions Editor : C. Janarthanan
Assistant Editors : B. Ramalakshmi, S. Revathi
Composition : M. Uma, Lissy John, N. Yamuna Devi
Cover Designer : M. Uma Maheswari
CIP Data : Prof. K. Hariharan

This book has been published in good faith that the work of the author is original. All efforts have been taken to make the material error-free. However, the author and publisher disclaim responsibility for any inadvertent errors.

To

My Parents, Teachers
and
Well Wishers

Foreword

New textbooks in crystallography don't appear with a great regularity although the field of X-ray crystallography is rapidly expanding. Many other fields are well represented by excellent texts where new or revised editions are published often to keep up with progress in research. The scope of the subject of X-ray crystallography is growing fast beyond traditional centres into biological domains. It is not far from now that it will be taught widely in almost all the scientific institutions. The demand for new texts in crystallography is enormous. Professor D. Velmurugan is one of India's outstanding crystallographers, who has contributed enormously both in theory and applications. He has an excellent international exposure, having worked in several foreign institutes. He is the most eminently suited person as an author for preparing a new textbook in crystallography. Indeed he has responded to the demand and has written a new book based on his lectures and seminars. The contents of this book cover the basic theory on crystals, nature of bonding in solids, crystal structure determination including various experimental techniques. The topics covered in this book are well tested through a number of questions and by providing solved problems at the end of the book. The book presents a means for trying together different methods and includes real numbers to give a better understanding of results. The contents of this book should serve the needs of both students and research scholars of crystallography. It may also stimulate the creation of a new course in institutions where crystallography is not yet being offered. Finally, I am sure that it should prove useful both to new crystallographers and to old hands.

Tej.P.Singh Ph.D., F.N.A., F.A.Sc., F.N.A.Sc., F.T.W.A.S.
Distinguished Biotechnologist
All India Institute of Medical Sciences
New Delhi

Preface

Crystallography has become an interdisciplinary subject and deals with the structure of the materials and three-dimensional arrangement of atoms in molecules. More than four lakh small molecular structures are now available from various databases. This book is intended to be a study material for the graduate and post-graduate students who study physics, chemistry, chemical physics and material science.

Since crystallography is a vast subject, it is not possible to cover all aspects of crystallography in this book. This book is considered as the first part of a series of crystallography textbooks I am planning to write.

An introduction is given first to the readers about the Solid State, Crystal Lattice, Unit Cell, Crystal Systems and Bravais Lattice. The second chapter begins with Crystal Geometry wherein crystal planes and Miller Indices are detailed. Typical crystal structures are presented. Coordination number and packing factor are then discussed. Symmetry elements, point groups and space groups are later detailed. A more detailed chapter on Bonding in Solids then begins. Various types of Bonding in Solids like covalent, ionic, metallic, hydrogen bonds and van der Waals bonds are discussed in detail. The detailed chapter dealing with the actual subject of crystallography then begins. The concept of Reciprocal Lattice and Ewald sphere are introduced first. Diffraction of X-rays by a crystal lattice is then detailed. Laue Diffraction is discussed and the experimental methods in X-ray Crystallography like Oscillation photograph method, Precession method, Debye–Scherrer powder method are detailed along with their applications. Experimental determination of structure factor amplitudes is then discussed. The method of locating an atom, namely, the computation of electron density is then discussed. The bottleneck "phase problem" is detailed and use of "Direct methods" in overcoming the phase problem is elaborately discussed. The software which uses this procedure is also detailed. Finally, neutron and electron diffraction are also discussed.

To appreciate the role of crystallography in determining three-dimensional crystal structures of molecules, the computer package SHELX with relevant plotting routines, has been elaborately dealt with.

Most chapters are provided with worked examples, and some problems (along with answers) have also been provided.

This book has been written, keeping in mind the students who encounter this new discipline for the first time. It would be useful to those writing competitive exams like GATE, NET, SLET and other tests.

Certain aspects of crystallography like electron diffraction and neutron diffraction are only briefly dealt with as these are specialized subjects and separate textbooks can be written for these topics.

My sole aim of writing this book is to enable graduate, postgraduate and research students to master the subject. I, along with my friends had convened many workshops/ seminars for this purpose. Special practical training in 3D crystal structure determination can also be given to interested readers.

D. Velmurugan

Acknowledgements

I wish to acknowledge the hard work and dedication of my research student Miss K.N.Vennila and I thank her whole-heartedly for her involvement in bringing out this book in a short span of time.

I thank Miss D. Gayathri, CSIR Senior Research Fellow, working with me for her Ph.D. programme, for the careful and critical reading of the manuscript and also for her valuable suggestions.

I thank Prof. T. P. Singh, Department of Biophysics, All India Institute of Medical Sciences, New Delhi for writing the foreword.

I thank MJP Publishers for their patience with me in bringing out this textbook as there were many occasions where I delayed submission of chapters due to my unavoidable academic commitments. I thank MJP Publishers for their interest in this book and their ready acceptance and cooperation when I suggested this topic.

I welcome suggestions/criticisms from the readers as this book has been written in a short notice and I will certainly take care of improving the book in the second edition.

D. Velmurugan

Contents

INTRODUCTION

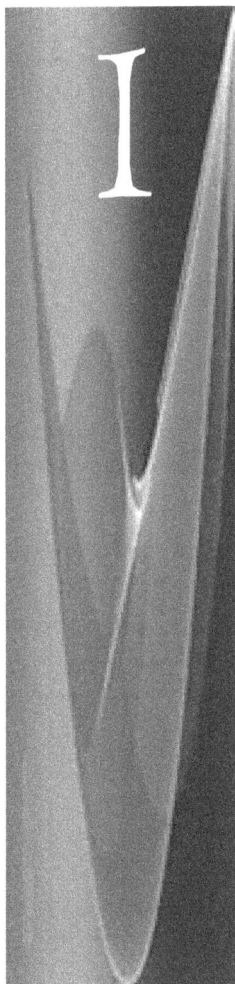

Matter is made up of atoms or molecules. The arrangement of these particles determines the state of matter. There are four recognized states of matter: solid, liquid, gas and plasma. To understand the different states in which matter can exist, we need to understand something called the **Kinetic Molecular Theory of Matter**. One of the basic concepts of the theory states that atoms and molecules possess energy of motion that we perceive as temperature. In other words, atoms and molecules are constantly moving, and we measure the energy of these movements as the temperature of the substance. The more energy a substance has, the more molecular movement there will be, and the higher the perceived temperature will be. An important point that follows this is that the amount of energy that atoms and molecules have (and thus the amount of movement) influences their interaction with each other. Unlike simple billiard balls, many atoms and molecules are attracted to each other as a result of various intermolecular forces such as hydrogen bonds, van der Waals forces, and others. Atoms and molecules that have relatively small amount of energy (and movement) will interact strongly with each other, while those that have relatively high energy will interact only slightly, if at all, with others.

How do these produce different states of matter? Atoms that have low energy interact strongly and tend to "lock" in place with respect to other atoms. Thus, collectively, these atoms form a hard substance, what we call as a solid. Atoms that possess high energy will move past each other freely, flying about a room, and forming what we call a gas. As it turns out, there are several known states of matter such as solid, liquid and gas.

THE SOLID STATE

Solids have a fixed volume and shape. In a solid, the atoms (or molecules) are in fixed positions relative to one another. They vibrate but stay in relative position. When the solid is heated the atoms vibrate faster. This causes the solid to grow slightly in size. It is said to **expand**. All solids expand on heating, especially metals.

If the atoms are arranged in a regular sequence, a **crystal** results. Metals and salts are crystalline. If the atoms are arranged haphazardly, the solid is said to be **amorphous** (Greek for **without shape**). Glass is a good example of an amorphous solid. Metals are crystalline because their atoms are arranged in a regular sequence. Amorphous solids are homogeneous and isotropic because there is no long-range order or periodicity in their internal atomic arrangement. By contrast, the crystalline state is characterized by a regular arrangement of atoms over large distances. Crystals are therefore anisotropic, their properties vary with direction. For example, the interatomic spacing varies with orientation within the crystal, as does the elastic response to an applied stress.

The outstanding macroscopic properties of crystalline solids are rigidity, incompressibility and characteristic shape. All crystalline solids are composed of orderly arrangement of atoms, ions, or molecules. The macroscopic result of the microscopic arrangements of the atoms, ions or molecules is exhibited in the symmetrical shapes of the crystalline solids.

The beauty and symmetry of crystals found all over the earth's crust have been appreciated by man even from early days. But in olden days, crystals were appreciated mainly for their ornamental value. Recently, however, new applications for single crystals have been discovered in solid-state devices. New applications continue to be found. For example, crystals are playing an important role in the development of high temperature superconductors. In the field of optics, new materials like titanium-doped sapphire are being developed as tunable lasers, which promise longer life, and more stable output than the present day lasers. The most exciting and at the same time widely anticipated contribution to molecular biology made by crystallographers in the recent years has been the direct visualization of atomic resolution of nucleic acid and variety of proteins with which it interacts.

A **single crystal**, also called monocrystal, is a crystalline solid in which the crystal lattice of the entire sample is continuous and unbroken to the edges of the sample, with no grain boundaries. Ideally, single crystals are free from internal boundaries. They give rise to a characteristic X-ray diffraction pattern. The **polycrystalline** sample is made up of a number of smaller crystals known as **crystallites**. Almost all common metals, and many ceramics are polycrystalline. The crystallites are often referred to as grains.

LATTICE

A regular, infinite arrangement of points in which every point has the same environment as any other point is known as lattice.

A crystal comprises a **space lattice** and a **basis** (Figure 1.1).

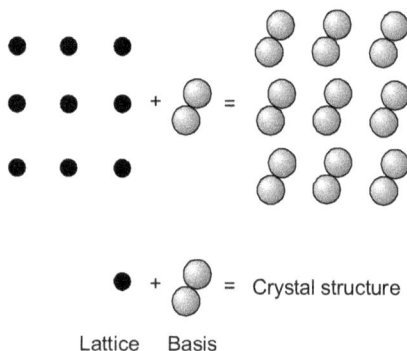

Figure 1.1   Lattice and basis

This type of regular periodic arrangement of atoms, ions or molecules in three-dimensional space is known as **crystal lattice**.

There are four important points about crystal lattices that are noteworthy for the study of crystals.

1. *Crystal faces develop along planes defined by the points in the lattice.* In other words, all crystal faces must intersect atoms or molecules that make up the points. A face is more commonly developed in a crystal if it intersects a larger number of lattice points. This is known as the Bravais Law.

For example, in the plane lattice shown in the Figure 1.2, faces will be more common if they develop along the lattice planes labelled 1, somewhat common if they develop along those labelled 2, and less and less common if they develop along planes labelled 3, 4, and 5.

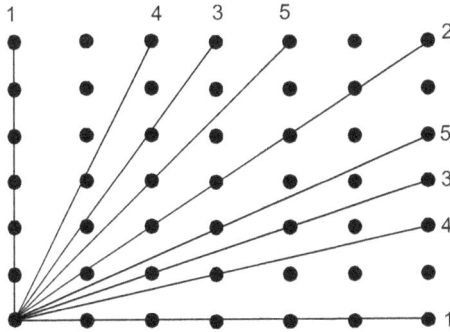

Figure 1.2 Lattice planes

2. *The angle between crystal faces is controlled by the spacing between lattice points.* As we can see from the imaginary 2-dimensional crystal lattice shown in the Figure 1.3a, the angle θ between the face that runs diagonally across the lattice and the horizontal face will depend on the spacing between the lattice points. Angles between faces are measured as the angle between the normals (lines perpendicular) to the faces. This applies in 3-dimensions as well.

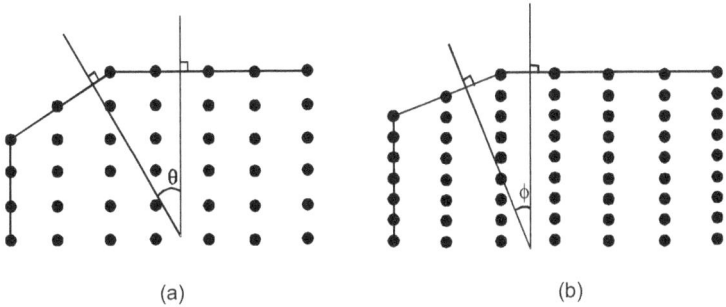

(a)   (b)

Figure 1.3 Imaginary 2-dimensional crystal lattices

Changing the lattice spacing changes the angular relationship. The lattice shown in the Figure 1.3b has the same horizontal spacing between lattice points, but a smaller vertical spacing. Note how the angle ϕ between the diagonal face and the horizontal face in this example is smaller than in Figure 1.3a.

3. *Since all crystals of the same substance will have the same spacing between lattice points (they have the same crystal structure), the angle between corresponding faces of the same compound will be the same.* This is known as the **Law of constancy of interfacial angles**.

4. *The symmetry of the lattice will determine the angular relationships between crystal faces.* Thus, in imperfect crystals or distorted crystals where the lengths of the edges are not equal, the symmetry can still be determined by the angles between the faces.

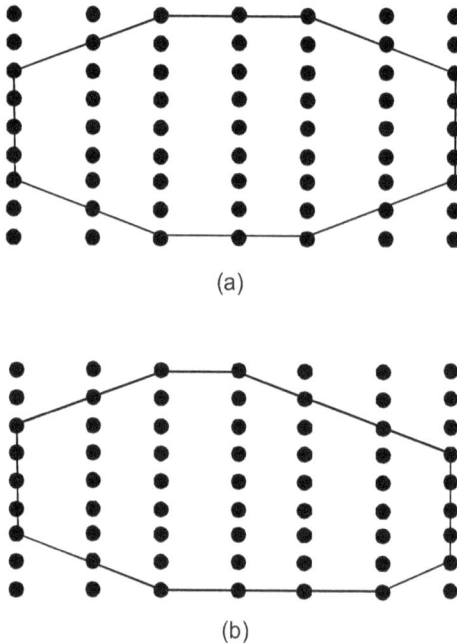

(a)

(b)

Figure 1.4 (a) Perfect crystal (b) Distorted crystal with distorted faces

Figure 1.4a shows a perfect crystal with the symmetrically related faces having equal lengths. Figure 1.4b shows a crystal built on the same lattice, but with distorted faces. Note that the angles between faces in the distorted crystal are the same as in the perfect crystal.

CRYSTALLOGRAPHIC AXES

In order to know which faces on different crystals are the corresponding faces, we need some kind of standard coordinate system onto which we can orient the crystals and thus be able to refer to different directions and different planes within the crystals. Such a coordinate system is based on the concept of the crystallographic axes.

The crystallographic axes are imaginary lines that we can draw within the crystal lattice. These will define a coordinate system within the crystal. For 3-dimensional space lattices we need 3, or in some cases 4, crystallographic axes that define directions within the crystal lattices. Depending on the symmetry of the lattice, the directions may or may not be perpendicular to one another, and the divisions along the coordinate axes may or may not be equal along the axes. As we will see later, the lengths of the axes are in some way proportional to the lattice spacing along an axis and this is defined by the smallest group of points necessary to allow for translational symmetry to reproduce the lattice.

In the following sections we will discuss the basic concepts of the crystallographic axes. As we will see, the axes are defined based on the symmetry of the lattice and the crystal. Each crystal system has different conventions that define the orientation of the axes, and the relative lengths of the axes.

A space lattice is a set of points such that a translation from any point P in the lattice by a vector $\boldsymbol{R}_{lmn} = l\boldsymbol{a} + m\boldsymbol{b} + n\boldsymbol{c}$ (l, m, n are integers) locates an exactly **equivalent point**, i.e., a point with the same environment as P. The vectors $\boldsymbol{a}$, $\boldsymbol{b}$, $\boldsymbol{c}$ are known as **lattice vectors**. In the Figure 1.5, two-dimensional lattice vector is given by $R = 2a + 3b$ which on translation yields a similar point in the space lattice.

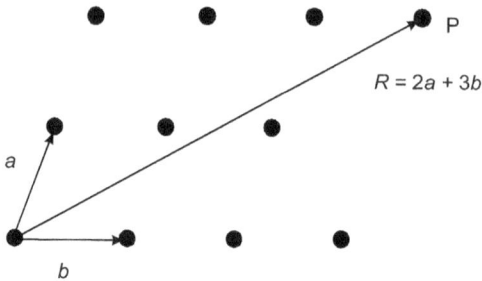

Figure 1.5   Lattice vector

The unit cell is defined as the parallelopiped (or parallelogram, in 2-D) whose edges are formed by these lattice vectors *a*, *b*, *c*. *It is the smallest repetitive unit in three dimensions, which on translation builds up the entire crystal structure.* Thus by considering a unit cell on its own, we automatically know all about the rest of the lattice. The unit cell is characterized by:

i. Three vectors (*a*, *b*, *c*) that form the edges of a parallelopiped;

ii. The angles between the vectors (alpha (α), the angle between *b* and *c*; beta (β), the angle between *a* and *c*; gamma (γ), the angle between *a* and *b*). These six parameters are called unit cell parameters (Figure 1.6).

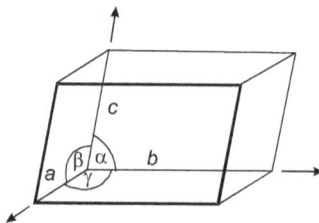

Figure 1.6   Unit cell

For each space lattice we can select **primitive lattice vectors** and an associated **primitive cell** containing just one lattice point (it does not matter where the lattice point is in the unit cell. It is conventionally placed at the corner). The unit cell that contains more than one lattice point is known as **non-primitive cell**. Note that a set of primitive vectors is not necessarily unique (Figure 1.7).

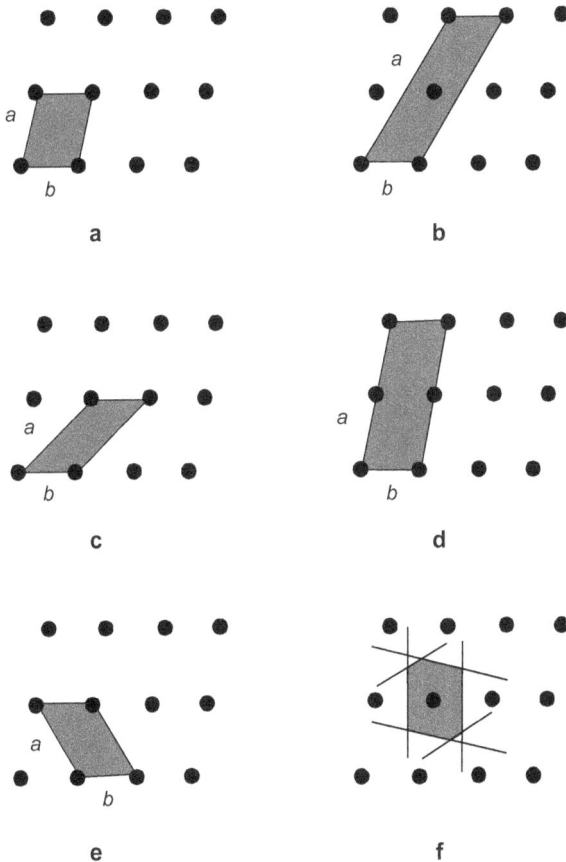

Figure 1.7   Primitive and non-primitive unit cells

The unit cells labelled in the Figure 1.7, **a**, **c**, **e** and **f** are primitive as they only contain one lattice point. The cells labelled **b** and **d** contain two lattice points and so are non-primitive. Cell **f** is a special case in which the cell has been formed by drawing the perpendicular bisectors (lines in 2-D, planes in 3-D) of lines joining a chosen lattice point to all of its neighbours. The smallest polygon (polyhedron in 3-D) thus formed is called the **Wigner Seitz** primitive cell.

There are two types of lattices—the **Bravais** and the **non-Bravais lattices**. In a Bravais lattice, all lattice points are equivalent, and hence by necessity all atoms in the crystal are of the same kind. In a **non-Bravais lattice**, some of the lattice points are non-equivalent. The non-Bravais lattice may

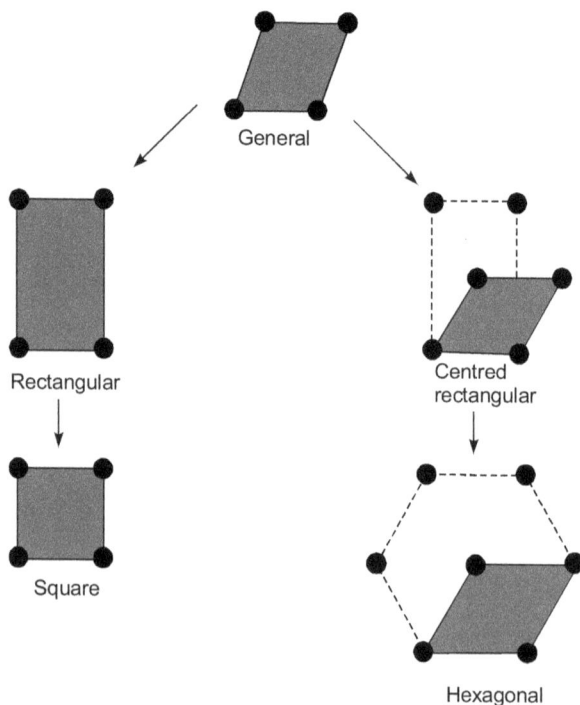

General

Rectangular

Square

Centred rectangular

Hexagonal

Figure 1.8   Two-dimensional Bravais lattices

be regarded as a combination of two or more interpenetrating Bravais lattices with fixed orientations relative to each other.

Depending on the crystal system, there are sometimes restrictions on the values that unit cell parameters can take. In 2-D, there are five Bravais lattices, which are shown in Figure 1.8.

There are more 3-D Bravais lattices than there are in 2-D, a total of 14 in all. These 14 lattices are subdivided into seven classes, with one lattice in each class being simple and the others being variations, which are described below.

SIMPLE LATTICE

The simplest cubic system is the simple cubic (sc) structure (Figure 1.9). The simple cubic system consists of one lattice point on each corner of the cube. Each lattice point is then shared equally between eight adjacent cubes, and the unit cell therefore contains in total one lattice point $\left(\dfrac{1}{8} \times 8\right)$. Given the radius of the atom or ion, the edge length is $a = 2\,r$ where r is the radius of the atom or ion. Polonium is the only metal reported to have this simple cubic structure.

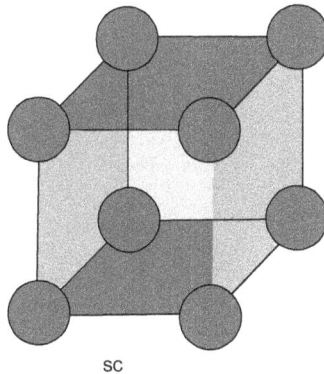

sc

Figure 1.9   Simple cubic unit cell

BODY CENTRED CUBIC

The body centred cubic (bcc) system has one lattice point in the centre of the unit cell in addition to the eight corner points. It has in total 2 lattice points per unit cell $\left(\frac{1}{8} \times 8 + 1\right)$. Each atom touches eight other host atoms along the body diagonal of the cube. The CsCl structure has a simple cubic arrangement of chloride ions with the cesium ion in the body centred hole (cubic hole) (Figure 1.10).

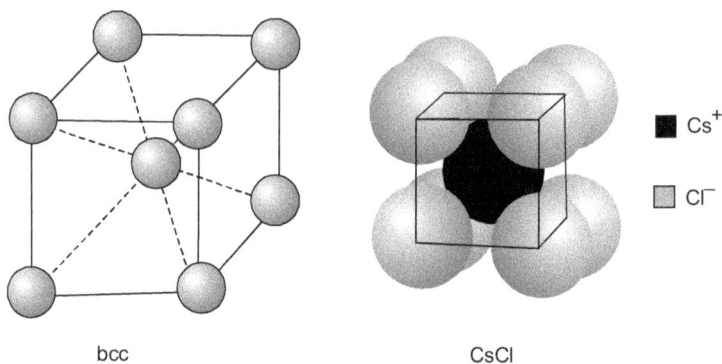

bcc   CsCl

■ Cs$^+$

□ Cl$^-$

Figure 1.10 bcc unit cell and CsCl

FACE CENTRED CUBIC

The face centred cubic (fcc) lattice has lattice points on the faces of the cube, giving a total of 4 lattice points $\left(\frac{1}{8} \times 8 + \frac{1}{2} \times 6\right)$. Each fcc atom is shared by two unit cells, so contribution from each fcc atom to each unit cell is only $\frac{1}{2}$. Since 6 such faces are there, the total contribution of all 6 fcc atoms will be $\frac{1}{2} \times 6 = 3$. ZnS, or zinc blende, has a face centred cubic arrangement of sulphide ions with zinc ions in every other tetrahedral hole (Figure 1.11). Examples of this structure includes KCl, PbS, MgO, MnO, KBr.

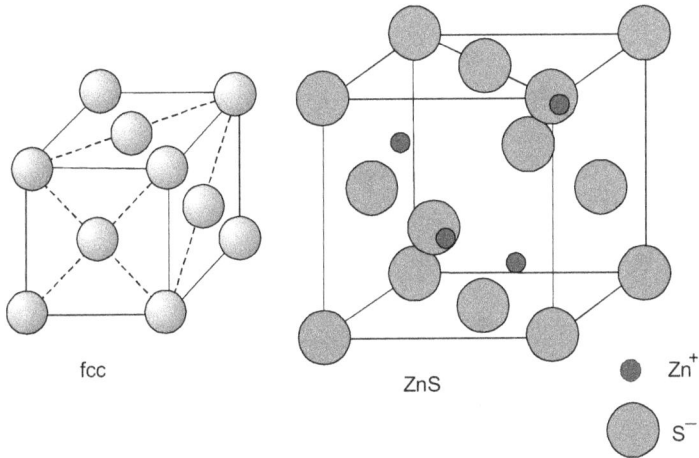

Figure 1.11 fcc unit cell and ZnS

The fourteen Bravais lattices are given in Figure 1.12 and are detailed in Table 1.1.

Table 1.1 The 14 Bravais lattices

Crystal system	Class of lattice	Parameters
Triclinic	Simple, P	$a \neq b \neq c$
		$\alpha \neq \beta \neq \gamma \neq 90°$
Monoclinic	Simple, P	$a \neq b \neq c$
	Base centred, C	$\alpha = \gamma = 90°$, $\beta > 90°$
Orthorhombic	Simple, P	$a \neq b \neq c$
	Body centred, I	$\alpha = \beta = \gamma = 90°$
	Face centred, F	
	Base centred, C	
Tetragonal	Simple, P	$a = b \neq c$
	Body centred, I	$\alpha = \beta = \gamma = 90°$
Rhombohedral (Trigonal)	Simple, P	$a = b = c$
		$\alpha = \beta = \gamma \neq 90°$
Hexagonal	Simple, P	$a = b \neq c$
		$\alpha = \beta = 90°$, $\gamma = 120°$
Cubic	Simple, P	$a = b = c$
	Body centred, I	$\alpha = \beta = \gamma = 90°$
	Face centred, F	

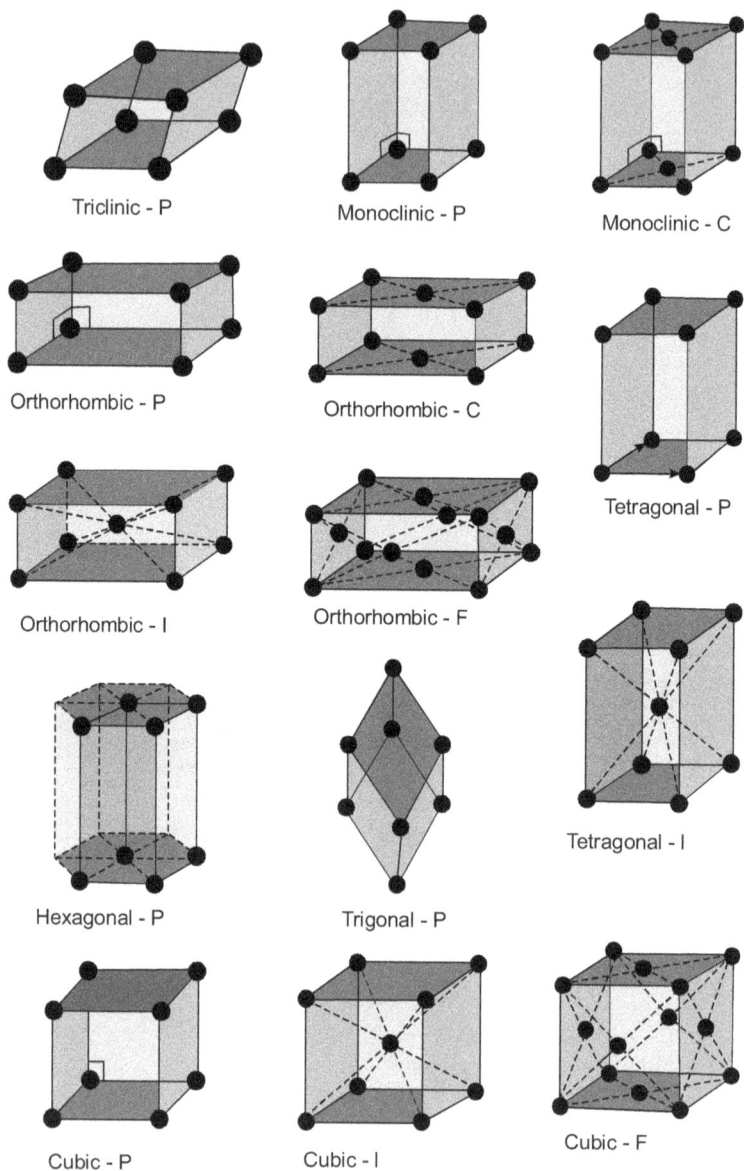

Triclinic - P

Monoclinic - P

Monoclinic - C

Orthorhombic - P

Orthorhombic - C

Tetragonal - P

Orthorhombic - I

Orthorhombic - F

Hexagonal - P

Trigonal - P

Tetragonal - I

Cubic - P

Cubic - I

Cubic - F

Figure 1.12 Bravais lattices

REVIEW QUESTIONS

1. Define space lattice.

2. Differentiate between primitive and non-primitive lattices.

3. Define Bravais and non-Bravais lattices.

4. Give two examples for orthorhombic and tetragonal crystal structures.

CRYSTAL GEOMETRY

MILLER INDICES

The orientation of a surface or a crystal plane may be defined by considering how the plane (or indeed any parallel plane) intersects the main crystallographic axes of the solid. The application of a set of rules leads to the assignment of the **Miller Indices**, (*hkl*)—*a set of numbers which quantify the intercepts and thus may be used to uniquely identify the plane or surface.*

HOW TO FIND MILLER INDICES?

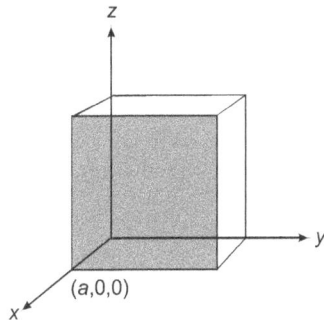

Figure 2.1   Unit cell with (*a*,0,0) intercepts

Step 1: *Identify the intercepts on the x-, y- and z- axes.* In Figure 2.1, the

intercept on the *x*-axis is at $x = a$ [at the point $(a, 0, 0)$] but the surface is parallel to the *y*- and *z*-axes—strictly therefore there is no intercept on these two axes but we shall consider the intercept to be at infinity (∞) for the special case where the plane is parallel to an axis. The intercepts on the *x*-, *y*- and *z*-axes are thus a, ∞, ∞.

Step 2: *Specify the intercepts in fractional coordinates.* Coordinates are converted to fractional coordinates by dividing by the respective cell-dimension. For example, a point (x, y, z) in a unit cell of dimensions $a \times b \times c$ has fractional coordinates of $(x/a, y/b, z/c)$. In the case of a cubic unit cell, each coordinate will simply be divided by the cubic cell constant, a. This gives the fractional intercepts, a/a, ∞/a, ∞/a, i.e., 1, ∞, ∞.

Step 3: *Take the reciprocals of the fractional intercepts.* This final manipulation generates the Miller indices which (by convention) should then be specified without being separated by any commas or other symbols. The Miller indices are also enclosed within standard brackets (....) when one is specifying a unique surface such as that being considered here.

The reciprocals of 1 and ∞ are 1 and 0 respectively, thus yielding Miller indices: (100)

So the surface/plane illustrated in Figure 2.1 is the (100) plane of the cubic crystal. Likewise the planes (111) and (101) are shown in Figure 2.2.

(*hkl*)—parentheses designate a **crystal face** or a **family of planes** throughout a crystal lattice.

[*hkl*]—square brackets designate a direction in the lattice from the origin to a point; used to collectively include all the faces of a crystal whose intercepts (i.e., edges) are parallel to each other. These are referred to as crystallographic **zones** and they represent a direction in the crystal lattice. Examples are shown in Figure 2.3.

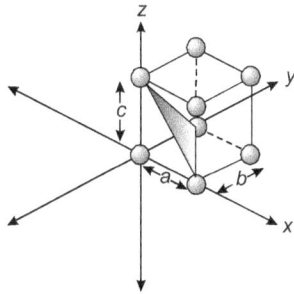

	a	b	c
Intercept length	1	1	1
Reciprocal	$\frac{1}{1}$	$\frac{1}{1}$	$\frac{1}{1}$
Cleared fraction	1	1	1
Miller indices		(111)	

(a)

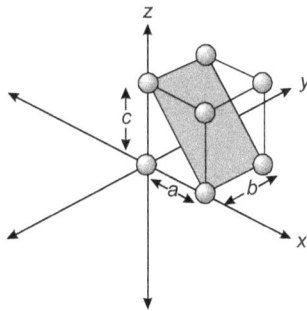

	a	b	c
Intercept length	1	∞	1
Reciprocal	$\frac{1}{1}$	$\frac{1}{\infty}$	$\frac{1}{1}$
Cleared fraction	1	0	1
Miller indices		(101)	

(b)

Figure 2.2 Miller indices. (a) For plane (111) (b) For plane (101)

{*hkl*}—"squiggley" brackets designate a set of face planes that are equivalent by the symmetry of the crystal. The set of face planes results in the **crystal form**. {100} in the isometric class includes (100), (010), (001), ($\bar{1}$00), (0$\bar{1}$0) and (00$\bar{1}$), while the triclinic {100} includes (100) plane only.

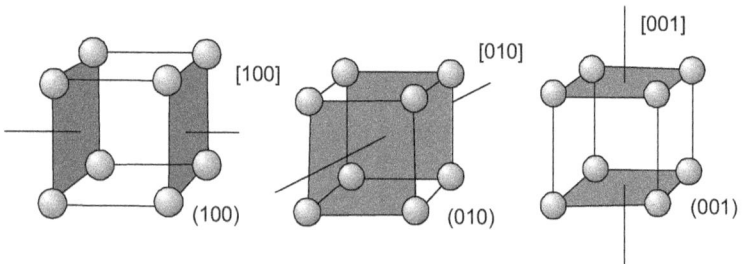

Figure 2.3   Planes and directions

<*hkl*>—equivalent directions

(*hkl*)—single plane

[*hkl*]—direction of a plane

{*hkl*}—set of parallel or equivalent planes

Some of the planes for simple cubic, body centred and face centred unit cells are shown in the Figure 2.4.

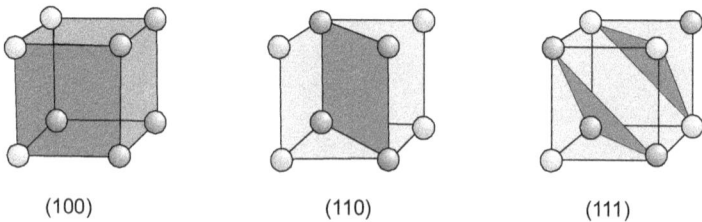

(100)   (110)   (111)

Figure 2.4   Crystal planes   (*Continues*)

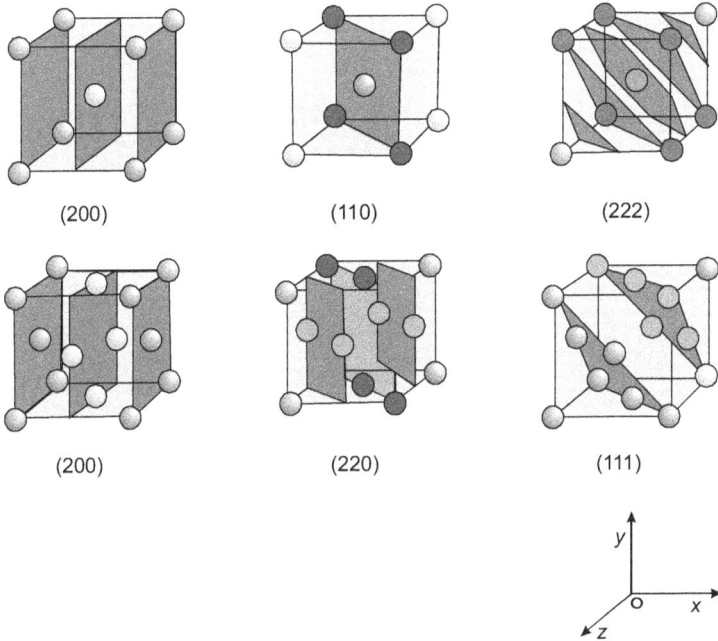

(200) (110) (222)

(200) (220) (111)

Figure 2.4 Crystal planes

Spacing between Planes of Miller Indices

In connection with X-ray diffraction from a crystal, one needs to know the interplanar distance between the parallel planes. Let us call this distance d_{hkl}. The actual formula depends on the crystal structure, and we confine ourselves to the case in which the axes are orthogonal. We can calculate this by referring the Figure 2.5, visualizing another plane parallel to the one shown and passing through the origin. The distance between these planes, d_{hkl}, is simply the length of the normal line drawn from the origin to the plane shown. Let us suppose that the angles which the normal lines make with the axes are α, β, γ and that the intercepts of the plane (hkl) with the axes are x, y, and z. Then it is evident from Figure 2.5 that

$$d_{hkl} = x \cos \alpha = y \cos \beta = z \cos \gamma \qquad (2.1)$$

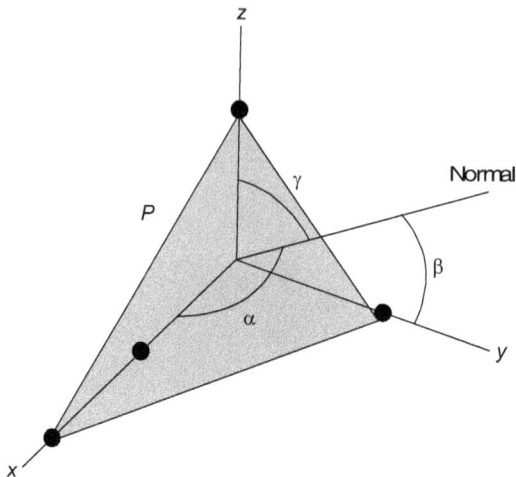

Figure 2.5   Interplanar spacing

But there is a relation between the directional cosines cos α, cos β, and cos γ. That is, $\cos^2 \alpha + \cos^2 \beta + \cos^2 \gamma = 1$. If we solve for cos α, cos β, and cos γ substituting into the one immediately above, and solve for d_{hkl} in terms of x, y, and z, we find that

$$d_{hkl} = 1/\sqrt{(1/x^2 + 1/y^2 + 1/z^2)} \qquad (2.2)$$

Now, x, y and z are related to the Miller indices h, k and l. If one reviews the process of defining these indices, one readily obtains the relations

$$h = n\,a/x, \quad k = n\,b/y, \quad l = n\,c/z \qquad (2.3)$$

where n is the common factor used to reduce the indices to the smallest integers possible. Solving for x, y, and z from equation 2.3 and substituting into equation 2.2, one obtains

$$d_{hkl} = \frac{n}{\left[\dfrac{h^2}{a^2} + \dfrac{k^2}{b^2} + \dfrac{l^2}{c^2} \right]^{1/2}} \qquad (2.4)$$

which is the required formula. Thus the interplanar distance of the (111) planes in a simple cubic crystal is

$$d = na/\sqrt{3}, \text{ where } a \text{ is the cubic edge.}$$

In the case of cubic lattice, there are 3 lattice types. The ratios $\dfrac{1}{d_{100}} : \dfrac{1}{d_{110}} : \dfrac{1}{d_{111}}$ are different for each lattice type. These are $1 : \sqrt{2} : \sqrt{3}$ for a simple cubic, $1 : \dfrac{1}{\sqrt{2}} : \sqrt{3}$ for a bcc and $1 : \sqrt{2} : \dfrac{\sqrt{3}}{2}$ for a fcc type. So knowing the values of glancing angles ratio, the ratio of interplanar spacings can be determined even without knowing the wavelength of X-rays. From that ratio, the type of lattice can be identified.

TYPICAL CRYSTAL STRUCTURES

Crystal Structure of Sodium Chloride

The unit cell of sodium chloride is of the type fcc, and this is reflected in the shape of NaCl crystals. The unit cell can be drawn with either the Na^+ ions at the corners, or with the Cl^- ions at the corners. If the unit cell is drawn with the Na^+ ions at the corners, then Na^+ ions are also present at the centre of each face of the unit cell.

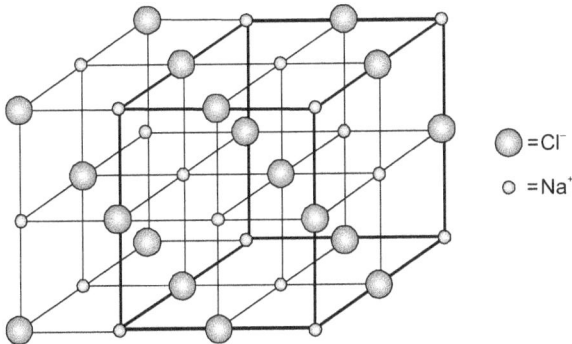

Figure 2.6   NaCl crystal structure

If the unit cell is drawn with the Cl^- ions at the corners, then Cl^- ions are also present at the centre of each face of the unit cell (Figure 2.6). Within the unit cell there must be an equal number of Na^+ and Cl^- ions. For the unit cell with the Cl^- ions at the centre of the faces, the top layer has $(1/8 + 1/8 + 1/8 + 1/8 + 1/2) = 1$ Cl^- ion, and $(1/4 + 1/4 + 1/4 + 1/4) = 1$ Na^+ ion. The middle layer has $(1/2 + 1/2 + 1/2 + 1/2) = 2$ Cl^- ions and $(1/4 + 1/4 + 1/4 + 1/4 + 1) = 2 Na^+$ ions. The bottom layer will contain the same as the top. The unit cell has a total of 4 Cl^- and 4 Na^+ ions in it. This equals the empirical formula NaCl.

Crystal Structure of Diamond

The diamond structure (Figure 2.7) has an fcc space lattice and a basis comprising two identical atoms. Alternatively, the structure may be viewed as two interpenetrating fcc lattices, one displaced relative to the other along a body diagonal by a quarter of its length. The atoms on each of the two fcc sub-lattices, have been drawn in dark and light grey to make the diagram clear, but in fact are identical. Relatively few elements exist in the diamond structure.

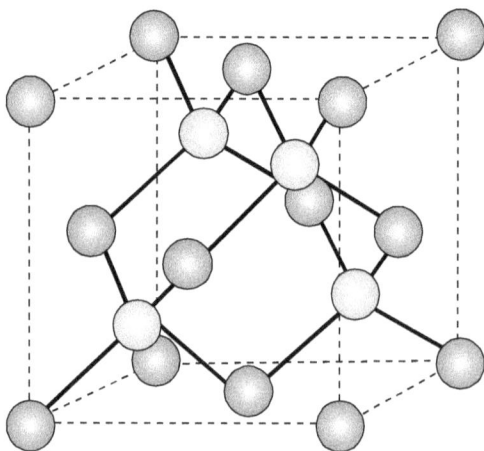

Figure 2.7 Diamond crystal structure

Carbon is the most obvious example, but the importance of this structure in solid-state physics is mainly due to the fact that this is the natural state of the semiconductors, silicon and germanium. Many two-element compounds form a diamond-like structure in which the first element occupies one fcc sub-lattice and the second element the other fcc sub-lattice. Another example of this type is ZnS whose crystal structure is shown in Figure 2.8.

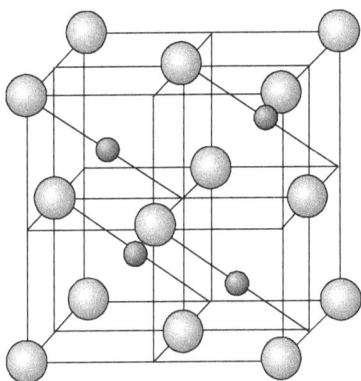

Figure 2.8 ZnS crystal structure

CRYSTAL PACKING

Many ions are spherical and many small molecules pack in a crystal lattice as essentially spherical entities. The packing of spheres is often used to model metal crystal structures. Spheres can pack in three dimensions in two general arrangements, the hexagonal close (hcp) and the face-centred close packing (fcc or ccp). The hcp packing has an ABABAB... two-layer sequence, where A and B represent the location of the layers. In other words, the third layer is exactly above the first layer in hcp. The fcc packing has a three-layer sequence, ABCABC... rather than the two-layer sequence of the hcp. The two types of packing are shown in Figure 2.9.

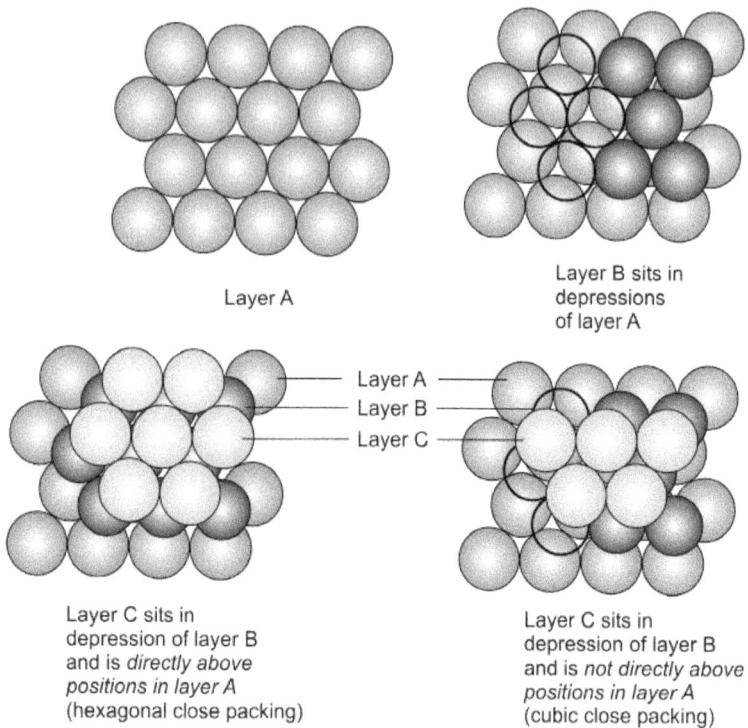

Layer A

Layer B sits in depressions of layer A

Layer A ——
Layer B ——
Layer C ——

Layer C sits in depression of layer B and is *directly above* positions in layer A (hexagonal close packing)

Layer C sits in depression of layer B and is *not directly above* positions in layer A (cubic close packing)

Figure 2.9   hcp and ccp

Hexagonal Close Packing

The hcp structure is based on the simple hexagonal lattice and has a two-atom basis. The second atom of the basis is positioned halfway up the cell and so forms another hexagonal layer of close-packed atoms halfway between the top and bottom layers of the unit cell (Figure 2.10). There is no way of choosing a primitive unit cell such that the basis contains only one atom. The primitive cell usually drawn for the hcp crystal structure is shown in bold. Hexagonal close packing in 3-D is shown in Figure 2.11. It is conventional to refer to the lattice constant (interatomic separation) in the plane of hexagons as a and that in the perpendicular, or axial, direction as c (where the latter is the height of the unit cell, and not the separation of adjacent layers of hexagonal planes). This packing leads to the possibility of two unique structures, depending on how planes of 2-D closest packed spheres are layered. If every other layer is exactly the same then we have a so-called ABABA... structure (Figure 2.11).

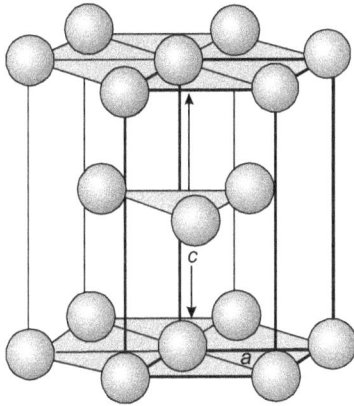

Figure 2.10 Hexagonal unit cell

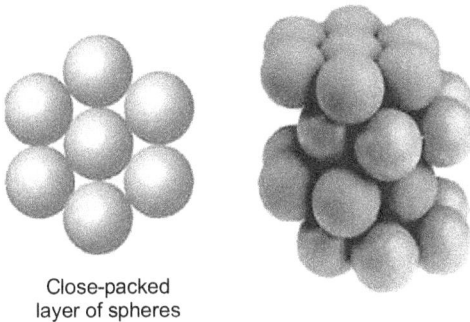

Close-packed
layer of spheres

Figure 2.11 hcp in three-dimension

Cubic Close Packing

The ABCABC structure is called face centred cubic (fcc) (Figure 2.12). It also has each atom with 12 nearest neighbours and the atoms fill 74.04% of the available space.

Figure 2.12   ccp in three-dimension

COORDINATION NUMBER

Crystals form because of the attraction between the atoms. Because they attract one another, it is often favourable to have many neighbours. Thus, the **coordination number**, or number of adjacent atoms, is important. Coordination number is defined as the total number of immediate neighbours of an atom in a unit cell. The coordination numbers for various crystal packing are shown in Figure 2.13.

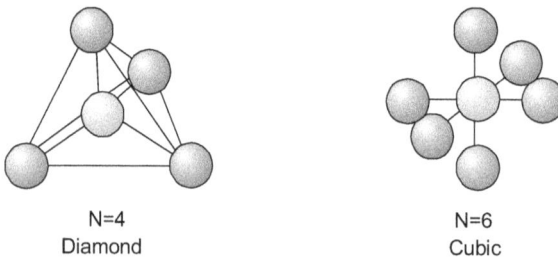

N=4   N=6
Diamond   Cubic

Figure 2.13   Coordination number   (*Continues*)

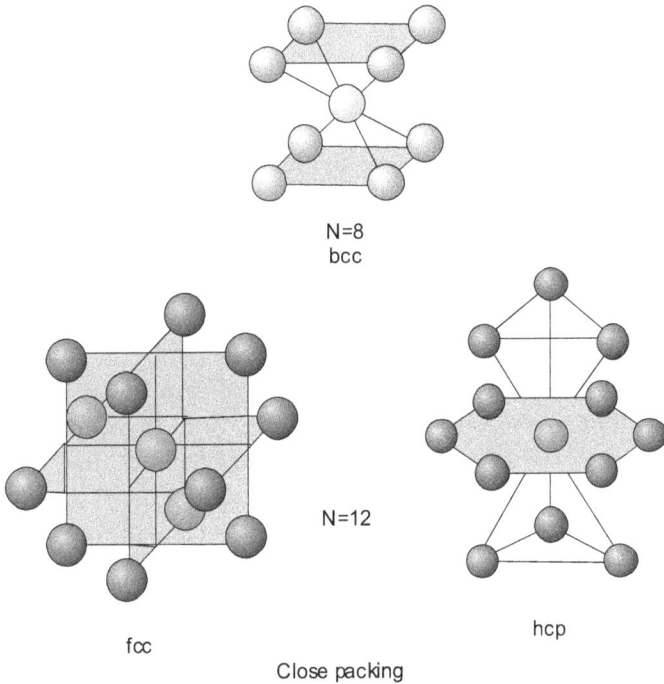

Figure 2.13 Coordination number

CRYSTAL PACKING FACTOR

In crystallography, **atomic packing factor** (sometimes called a **packing fraction**) is the fraction of volume in a crystal structure that is occupied by atoms. It is dimensionless and always less than unity. For practical purposes, the APF of a crystal structure is determined by assuming that atoms are rigid spheres. It is represented mathematically as

$$APF = \frac{N_{atoms} V_{atom}}{V_{crystal}}$$

where,

   N is the number of atoms in the crystal and

   V is the volume.

It can be proven mathematically that one-component (one type of atom) close-packed structures that have the most dense arrangement of atoms, have an APF of 0.74. The empty spaces between the atoms are interstitial sites.

Also because the atoms attract one another, there is a tendency to squeeze out as much empty space as possible. The **packing efficiency** is the fraction of the crystal (or unit cell) actually occupied by the atoms. It must always be less than 100% because it is impossible to pack spheres (atoms are usually spherical) without having some empty space between them. In reality, this number can be higher, given specific intermolecular factors. For multiple-component structures, the APF can exceed 0.74.

SC

The simple cubic crystal structure contains eight atoms at each corner of the cube. Each atom touches its neighbour and hence the length of each side of the simple cubic structure is $2r$, where r is the radius of the atom. Therefore,

$$APF = \frac{N_{atoms}V_{atom}}{V_{crystal}}$$

$$= \frac{(4/3)\pi r^3}{(2r)^3}$$

$$= \frac{\pi}{6}$$

$$= 0.52$$

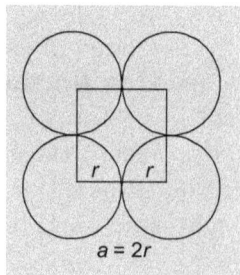

$a = 2r$

bcc

The body centred cubic crystal structure contains eight atoms at each corner of the cube and one atom at the centre. Because the volumes of the corner atoms are shared between adjacent cells, each bcc crystal contains only two whole atoms.

Each corner atom touches the centre atom. A line that is drawn from one corner of the cube through the centre and to the other corner passes through $4r$, where r is the radius of an atom. By geometry, the length of the diagonal is $\sqrt{3}a$. Therefore, the length of each side of the bcc structure can be related to the radius of the atom, i.e.,

$$a = \frac{4r}{\sqrt{3}}$$

Knowing this and the formula for the volume of a sphere, it becomes possible to calculate the APF.

$$
\begin{aligned}
APF &= \frac{N_{atoms} V_{atom}}{V_{crystal}} \\
&= \frac{2(4/3)\pi r^3}{(4r/\sqrt{3})^3} \\
&= \frac{\pi\sqrt{3}}{8} \\
&= 0.68
\end{aligned}
$$

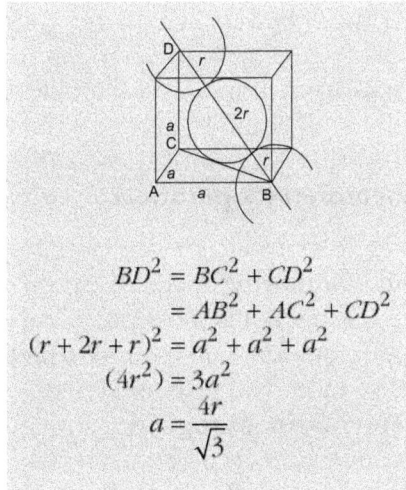

$$
\begin{aligned}
BD^2 &= BC^2 + CD^2 \\
&= AB^2 + AC^2 + CD^2 \\
(r + 2r + r)^2 &= a^2 + a^2 + a^2 \\
(4r^2) &= 3a^2 \\
a &= \frac{4r}{\sqrt{3}}
\end{aligned}
$$

fcc

There are four atoms per fcc unit cell, i.e., $N_{atom} = 4$, and the total fcc sphere volume is $V_{atom} = 4/3\pi r^3$. The total unit cell volume is $V_{crystal} = 16r^3\sqrt{2}$.

Therefore, the atomic packing factor is

$$APF = \frac{N_{atoms}V_{atom}}{V_{crystal}}$$

$$= \frac{4(4/3)\pi r^3}{(16r^3/\sqrt{2})}$$

$$= \frac{\pi}{3\sqrt{2}}$$

$$= 0.74$$

$$BC^2 = AB^2 + AC^2$$
$$(r + 2r + r)^2 = a^2 + a^2$$
$$16r^2 = 2a^2$$
$$r = \frac{a}{2\sqrt{2}}$$
$$a = 2\sqrt{2}\,r$$

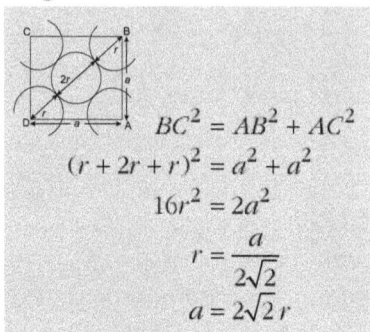

SYMMETRY OPERATIONS/ELEMENTS

In the most general description, **symmetry** can be described as the invariance of an object under some kind of transformation. In somewhat narrower terms, an object is described as symmetric with respect to an operation if the object does not appear to change after the transformation. In crystallography, most types of symmetry involve a movement of object such as some type of rotation or translation. The movement is called the **symmetry operation**.

A molecule or object is said to possess a symmetry if that symmetry operation when applied leaves the molecule unchanged. Each operation is performed relative to a point, line, or plane called a **symmetry element**. There are two common ways to designate symmetry operations, the **Hermann–Mauguin** convention and the **Schöenflies** convention. The Hermann–Mauguin system was developed specifically for describing crystals and crystallographic symmetry. The **Schöenflies** convention was conceived primarily to describe symmetry in optical spectroscopy and quantum mechanics.

There are **four types of symmetry operations**, which lead to superimposition of an object on itself: **rotation,**

translation, reflection and **inversion**. A **symmetry element** is a geometrical entity such as a point, line, or plane about which a symmetry operation is performed. The symmetry of any object can be described by some combination of these symmetry operations. The symmetry of any aggregate or crystal of a biological molecule is only described by rotation and/or translation operations. This is because, for example, biological protein molecules mainly consist of L-amino acids, hence, reflection or inversion symmetries are not allowed.

A **point group** is a collection of symmetry operations that define the symmetry about a point (at least one point remains invariant). Two systems of notations are used for point groups: i) the **S** or **Schöenflies notation** (capital letters; mainly used by spectroscopists) and ii) the **H–M** or **Hermann–Mauguin symbol** (an explicit list of the symmetry elements, commonly preferred by crystallographers). The four types of symmetries about a point are rotational symmetry, translational symmetry, mirror symmetry, inversion symmetry, and improper rotations.

Rotation (*n*)

An *n*-fold (C_n) proper rotation operation represents a movement of 360/*n* degrees around an axis through the object. If an *n* fold rotation operation is repeated *n* times, then the object returns to its original position.

Consider an equilateral triangle. This triangle contains a 3-fold rotation (C_3) axis in the centre of the triangle and perpendicular to the plane of the vertices of the triangle. By rotating the triangle by 360°/3 or 120°, one point of the triangle is made to coincide with another point.

Because of the inherent lattice nature of "crystalline" objects, only 1(E), 2(C_2), 3(C_3), 4(C_4), and 6 fold (C_6) (**Schöenflies** notation in parentheses) rotation operations are known. A 1-fold (E) rotation operation, which implies no movement of the object, is referred to as the identity operation.

Translation

If a pattern is reproduced by translation $T = nt$, where n is an integer, then t is a **translation symmetry element** of the pattern. A motif is an object or group of objects that is repeated by the translation symmetry elements of the lattice to construct the pattern (crystal).

Reflection (m)

Each point in the object is converted to an identical point by projecting through a mirror plane and extending an equal distance beyond this plane. The reflection symmetry is also known as mirror symmetry (Figure 2.14).

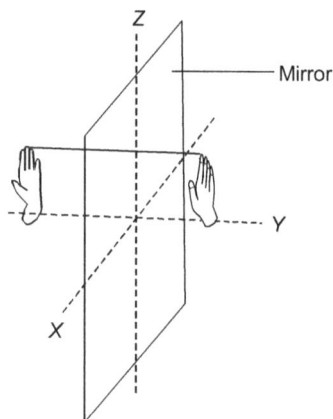

Figure 2.14   Mirror symmetry

Inversion Symmetry (*i*)

Each point in the object is converted to an identical point by projecting through a common centre and extending an equal distance beyond this centre. Objects with *i* symmetry are said to be centrosymmetric. If there is any atom in the position (x, y, z), its centrosymmetric position is in $(-x, -y, -z)$ (i.e., there is an identical atom at $-x, -y, -z$) (Figure 2.15).

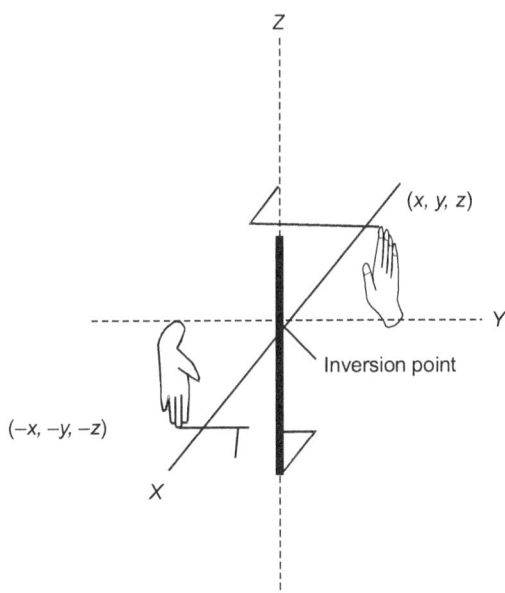

Figure 2.15 Inversion symmetry

There are 32 such possible number of combinations of these symmetry operations, which leads to 32-point groups in the 14 Bravais lattices.

We can combine the 14 Bravais lattices and the 32-point groups as shown in Table 2.1.

Table 2.1   Bravais lattices and 32-point groups

Crystal class	Bravais lattices	Point groups
Triclinic	P	1, $\bar{1}$
Monoclinic	P, C	2, m, 2/m
Orthorhombic	P, C, F, I	222, mm2, 2/m 2/m 2/m
Trigonal	P, I	3, $\bar{3}$, 32, 3m, $\bar{3}$ 2/m
Hexagonal	P	6, $\bar{6}$, 6/m, 622, 6mm, $\bar{6}$ m2, 6/m 2/m 2/m
Tetragonal	P, I	4, $\bar{4}$, 4/m, 422, 4mm, $\bar{4}$ 2m, 4/m 2/m 2/m
Cubic	P, F, I	23, 2/m$\bar{3}$ 432, $\bar{4}$ 3m, 4/m $\bar{3}$ 2/m

Space Groups

*The translational symmetry of a crystalline lattice needs to be combined with the point group symmetry of an object in order to represent the whole symmetry of the crystal, called the **space group symmetry**.* In addition to lattice translations, it is possible to combine proper rotation axes with translations of part of the unit cell to create screw axes. Similarly, mirror planes may be combined with partial translations of the cell to generate glide planes. There are 230 such possible combinations which are known to be in seven crystal systems. Screw axes and glide planes are similar to cell centring operations and simple cell translations in which one group of atoms transform into an entirely different (but to appearance, they are identical) group of atoms.

A **screw axis (rotation + translation)** occurs when a proper rotation axis operation is followed by a translation by a fraction of the unit cell in the direction of the rotation.

The symbol for a screw axis is n_m where n indicates the type of rotation and the translation is (m/n) of the unit cell. Thus a 3_1 screw axis is a 3 fold rotation followed by a translation of 1/3 of the unit cell. Performing this operation three times is equivalent to a full unit cell translation.

Glide planes (reflection + translation) occur when a mirror operation is followed by a translation of a fraction of the unit cell parallel with the mirror plane. The glide directions are usually parallel with a unit cell direction or a combination of cell directions. When glide planes are described outside of the context of a particular space group, they are given the symbols f_g in which the letter **g** indicates the direction of the mirror type operation and f indicates the direction of translation. Thus an a_b, an a glide in the **b** direction, means that the object is reflected in a plane parallel with the (010) planes and then translated by $a/2$ of the unit cell in the **a** direction. Glide planes exist in all three directions and in pairs of directions. Glides that translate by half of the cell in two different directions are called n glide planes. An object undergoes an n_c operation when it is reflected in the (001) plane, and translated by $(a + b)/2$ in the **a + b** direction. Two of these types of glide operations are needed to bring about an operation that is equivalent to a unit cell translation.

There is one additional type of glide plane, the **diamond glide**, d. It occurs only in space groups with face- or body-centred cells, and is characterized by a translation of $(a + b)/4$, $(b + c)/4$, or $(c + a)/4$. As the denominator implies, 4 consecutive d glides are required to return an object to a lattice-translated version of it.

A **space group is designated** by a capital letter identifying the lattice type (P, A, F, etc.) followed by the point group symbol in which the rotation and reflection elements are extended to include screw axes and glide planes. Note that the point group symmetry for a given space group can be determined by removing the cell-centring symbol of the space group and replacing all screw axes by similar rotation axes and replacing all glide planes with mirror planes. The point group symmetry for a space group describes the true symmetry of its reciprocal lattice.

EQUIVALENT POSITIONS IN A UNIT CELL

Plane Groups

The application of some of group-theoretical results to real crystals requires appreciation of the constraints that symmetry elements impose on the entire space surrounding them. Thus, if an atom is placed at some point (x, y, z) in a unit cell, it will be repeated a number of times, determined by the aggregate of symmetry elements present and their disposition relative to the point (x, y, z). Another way of stating this is to realize that, in real crystals there are like atoms that are related to each other (made equivalent) by the symmetry of the crystal. The set of positions occupied by such symmetry-equivalent atoms is called an **equipoint** set of the crystal, and the number of the equivalent points in the set is called its rank. Finally, all the distinguishable sets that exist in one cell are called the equipoints of a space group.

Let us consider the plane group P4. Figure 2.16 shows a point having the general coordinates xy repeated by the symmetry of P4. It is usual to give the coordinates of a point in a unit cell in terms of fractions of the cell edges, thereby making them independent of actual cell dimensions. The value of the fraction x is the actual distance along the a axis, from the origin to the point, divided by the actual length of a. Similarly, the fraction y is the actual distance along the b axis divided by the length of b. Thus one complete translation along a is $a/a = 1$, etc. Proceeding in a clockwise manner, the coordinates of the next point in Figure 2.16 are the fraction y along the a axis and $1 - x = 1 + \bar{x} = \bar{x}$ along the b axis. Since it is customary to list the coordinate first along a and next along b, the coordinates of this point are $y\bar{x}$. The coordinates of the other equivalent points are also indicated in Figure 2.16. Because there are four such points related by the symmetry of P4, the rank of the equipoint xy is 4 in this plane group. Since each of the coordinates of this equipoint, x and y, is variable between 0 and 1, this position in the cell is called a general position.

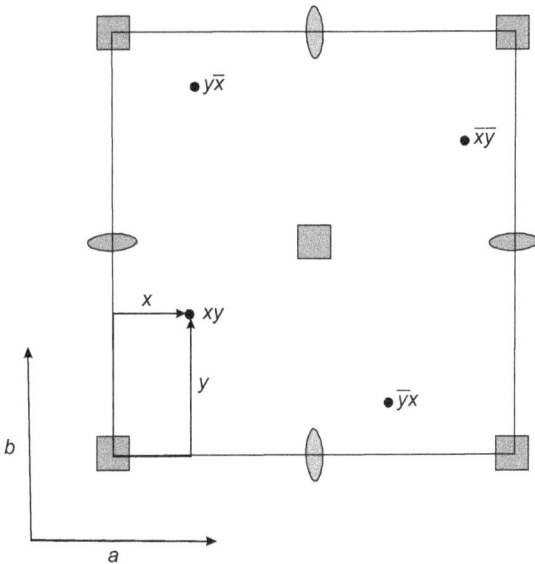

Figure 2.16 P4 symmetry

In the case of pairs of points related by the 2-fold axes in P4, these points coalesce to single points on each 2-fold, as indicated in Figure 2.17 by the dashed lines. The co-ordinates of these equivalent points (the black dots in the Figure 2.17) are ½ 0 and 0 ½, and there are two such points

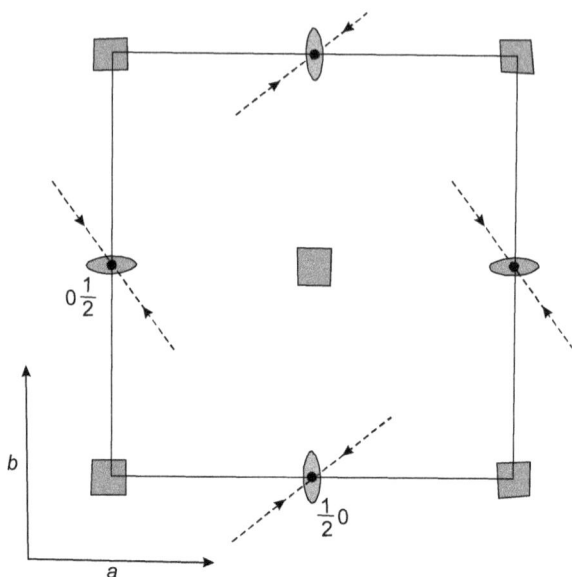

Figure 2.17   Special positions

in each cell. Finally the points on the two different 4-fold axes, at 00 and ½ ½, each having the rank 1. Since the x and y values of these equipoints are fixed, they are called **special positions**. Table 2.2 lists the various equipoints in P4 and their symmetry. For example, the equipoint 00 must have 4-fold symmetry since it occupies a 4-fold axis. Also note that the arithmetic product of the first two columns in

Table 2.2   Equipoints in P4

Rank of equipoint	Symmetry of location	Coordinates of equivalent points
1	4	00
1	4	½ ½
2	2	0 ½ , ½ 0
4	1	$xy, y\bar{x}, \bar{x}\bar{y}, \bar{y}x$

Table 2.2 is a constant. This is so because a point lying on a symmetry axis is not repeated by that axis; hence its rank is decreased by the fold of the axis. This relationship holds true whether the axis is a proper or an improper axis.

Space Groups

Let us consider the space group P4$_2$/m for deriving the sets of equivalent positions. Figure 2.18 shows the equivalent points generated by a fourfold rotation axis in the c direction; a counterclockwise rotation of 90° generates a point with coordinates $\bar{y}, x, z$ from a point with coordinates x, y, z. If for example, our initial point had coordinates 0.1, 0.2, 0.3, the generated point would be at –0.2, 0.1, 0.3 (which may be written as 0.8, 0.1, 0.3 because of the periodicity of the 4 arrangement). Application of the symmetry operator 4 to point xyz thus gives a point whose x coordinate is $-y$, whose y coordinate is x, and whose z

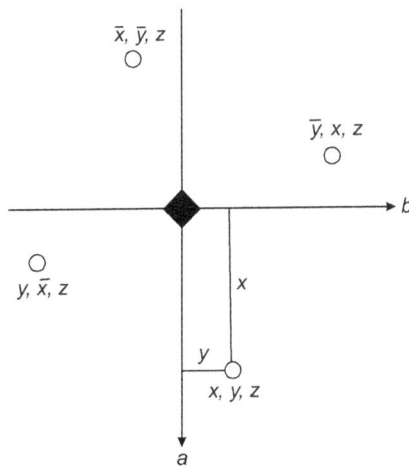

Figure 2.18 Equivalent points generated

coordinate is z. When 4 is applied to $\overline{y}, x, z$ the result is $\overline{x}, \overline{y}, z$ and still together application gives $y, \overline{x}, z$. One more application would give x, y, z again, so there are just four points related by the operation. The operator required in P4/m is 4_2 rather than 4. The difference is simply that a translation of $1/2c$ must be included in each operation, so the equivalent points generated are those in Figure 2.19. x, y, z; $\overline{y}, x, 1/2 + z$; $\overline{x}, \overline{y}, z$; $y, \overline{x}, 1/2 + z$. Space group P4$_2$/m

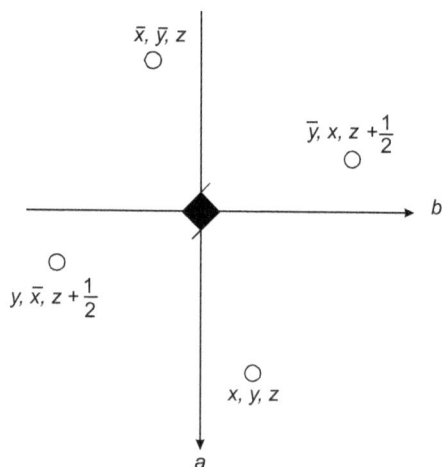

Figure 2.19 Equivalent points generated by symmetry operation 4_2

also includes a reflection perpendicular to the 4_2 axis. For each point x, y, z this reflection gives a point $x, y, \overline{z}$, so in addition to the four points generated by 4_2 we get four more points by changing the sign of z. Noting that $-(\frac{1}{2} + z) = -\frac{1}{2} - z$ is equivalent to $\frac{1}{2} - z$, we have $x, y, \overline{z}$; $\overline{y}, x, \frac{1}{2} - z$; $\overline{x}, \overline{y}, \overline{z}$; $y, \overline{x}, \frac{1}{2} - z$. These eight points constitute the general positions for space group P4$_2$/m.

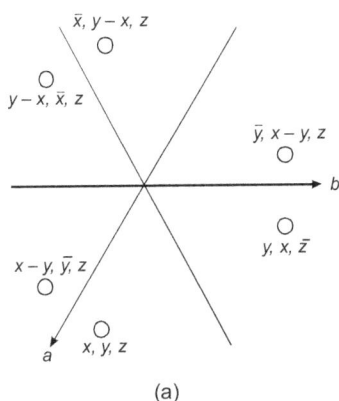

$\bar{x}, y - x, z$

$y - x, \bar{x}, z$

$\bar{y}, x - y, z$

b

$x - y, \bar{y}, z$

$y, x, \bar{z}$

a x, y, z

(a)

Figure 2.20   Equivalent points in P321

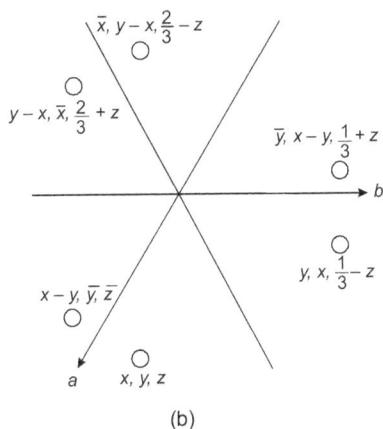

$\bar{x}, y - x, \frac{2}{3} - z$

$y - x, \bar{x}, \frac{2}{3} + z$

$\bar{y}, x - y, \frac{1}{3} + z$

b

$x - y, \bar{y}, \bar{z}$

$y, x, \frac{1}{3} - z$

a x, y, z

(b)

Figure 2.21   Equivalent points in P3₁21

Similarly the equivalent points generated by the symmetry operation P321 and P3₁21 are shown in Figures 2.20 and 2.21 respectively.

For further illustration let us consider the monoclinic space group P2₁ in which the primitive unit cell possesses twofold rotation followed by translation along *y*-axis. This combination of symmetry operations is known as screw axes. If an atom is placed at some point *xyz*, then the symmetry equivalent point will be $-x$, $y + \frac{1}{2}$, $-z$. Similarly for the space group P2$_1$/c, the special positions are given in Table 2.3.

Table 2.3   Special positions for P2$_1$/c

xyz	Multiplicity	Point symmetry
x y z	4	1
½ 0 ½	2	−1
½ 0 0	2	−1
0 0 ½	2	−1
0 0 0	2	−1

The equivalent positions are $-x$, $y + \frac{1}{2}$, $-z + \frac{1}{2}$ and $-x$, $-y$, $-z$ and x, $-y + \frac{1}{2}$, $z + \frac{1}{2}$.

REVIEW QUESTIONS

1. Define Miller indices and explain the steps in determining the Miller indices for the given intercepts.

   (a) 102   (b) 022   (c) 121   (d) ∞10

2. Deduce the formula to find the interplanar spacing in case of triclinic crystal system.

3. What is coordination number?

4. Explain the term "packing fraction".

5. Define point group.

6. What is meant by centro-symmetry?

7. Explain *a*, *b*, *c* and *n* glide.

8. Explain screw axis.

SOLVED PROBLEMS

Problem 1 Show that the maximum radius of the sphere that can just fit into the void at the body centre of the fcc structure coordinated by the facial atoms is $0.414r$, where r is the radius of the atom.

Solution

Let r be the radius of the atoms and R be the radius of the sphere that can just fit into the void.

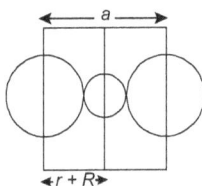

$$r + R = a/2$$
$$R = a/2 - r$$

For fcc structure,

$$a = 4r/\sqrt{2}$$

Thus
$$R = \frac{4r}{2\sqrt{2}} - r = \frac{2r}{\sqrt{2}} - r$$
$$R = r(\sqrt{2} - 1) = 0.414r$$
$$\therefore R = 0.414r \qquad \text{[Answer]}$$

Problem 2 Calculate the interplanar spacing distance of a cubic unit cell whose side is 2.1 Å for the [110] family of planes.

Solution

Interplanar distance for a cubic system
$$d = \frac{a}{\sqrt{h^2 + k^2 + l^2}}$$

$$d_{110} = \frac{2.1 \times 10^{-10}}{\sqrt{1^2 + 1^2 + 0^2}}$$

$$= \frac{2.1}{\sqrt{2}} \times 10^{-10} = 1.49 \text{ Å}$$

$$\therefore d_{110} = 1.49 \text{ Å} \qquad \text{[Answer]}$$

Problem 3 In an orthorhombic crystal, a lattice plane cuts intercepts of lengths $2a$, $-b$ and $3c/2$ along three axes. Deduce the Miller indices of the plane. a, b, c are primitive vectors of the unit cell.

Solution

$ra : sb : tc = 2a : -b : 3c/2$

$r : s : t = 2 : -1 : 3/2$

$\therefore \quad 1/r : 1/s : 1/t = 1/2 : -1/1 : 2/3$

Multiplying by 6, $1/r : 1/s : 1/t = 3 : -6 : 4$

$\therefore$ Miller indices of the plane is $(3\bar{6}4)$ \qquad [Answer]

Problem 4 Calculate the volume of a monoclinic unit cell with cell parameters $a = 14.84$ Å, $b = 11.19$ Å, $c = 16.09$ Å and $\alpha = 90°$, $\beta = 112.48°$ and $\gamma = 90°$.

Solution

Unit cell volume of monoclinic crystal system $= V = abc \sin \beta$

$V = 14.84 \times 11.19 \times 16.09 \times \sin 112.48$

$= 2671.9 \times \sin (112.48)$

$= 2468.87$ Å^3

Volume of the given monoclinic unit cell is 2469 Å^3

[Answer]

Problem 5 Deduce the Miller indices for the plane a, b, c whose intercepts on x, y and z axis are 1, -3 and ∞.

Solution

$$
\begin{array}{llll}
\text{Intercepts} & : & 1 & -3 & \infty \\
\text{Reciprocal} & : & 1/1 & -1/3 & 1/\infty \\
& & 1 & -1/3 & 0 \\
\therefore \ \text{Miller indices} & : & 1 \times 3 & -1/3 \times 3 & 0 \times 3 \\
& & 3 & -1 & 0
\end{array}
$$

Miller indices for the plane a, b, c is ($3\ \bar{1}\ 0$) [Answer]

Problem 6 Find the lattice constant for a simple cubic unit cell whose (110) interplanar spacing is 1.56 Å.

Solution

Given d = 1.56 Å, (hkl) = (110)

$$d = \frac{a}{\sqrt{h^2 + k^2 + l^2}}$$

$$\therefore \ a = d \times \sqrt{h^2 + k^2 + l^2}$$

$$= 1.56 \times 10^{-10} \times \sqrt{1^2 + 1^2 + 0^2}$$

$$= 1.56 \times 10^{-10} \times \sqrt{2}$$

$$= 2.206 \times 10^{-10}$$

$\therefore$ The lattice constant a = 2.21 Å   [Answer]

Problem 7 Lead is fcc and its atomic radius is 0.175 nm. What is its unit cell edge length?

Solution

$$\text{Atomic radius } r = \frac{a\sqrt{2}}{4}$$

$$\therefore a = \frac{4r}{\sqrt{2}}$$

$$= \frac{4 \times 0.175 \times 10^{-9}}{\sqrt{2}}$$

$$= 0.495 \times 10^{-9}$$

The edge length of lead unit cell is 4.95 Å. [Answer]

EXERCISES

1. Zinc has hcp structure. The height of the unit cell is 0.392 nm. The nearest neighbour distance is 0.22 nm. The atomic weight of zinc is 65.37. Calculate the volume of the unit cell.

   [Answer: 49.3 Å^3]

2. In a simple cubic crystal, find the ratio of intercepts on the three axes by the (123) plane.

   [Answer: 6 : 3 : 2]

3. Calculate the interplanar spacing for (310) planes in a simple cubic lattice. The lattice constant is 3.4 × 10^{-10} m.

   [Answer: 1.075 Å]

4. Draw the plane with the following indices in a cubic unit cell.

   i. 111   ii. 100
   iii. 212   iv. 010

5. Calculate the volume of an orthorhombic unit cell with the cell parameters, a = 20.94 Å, b = 10.58 Å, c = 11.01 Å and $\alpha = \beta = \gamma = 90°$.

   [Answer: 2440.6 Å^3]

6. Calculate the unit cell volume of a triclinic crystal system with cell parameters, $a = 9.34$ Å, $b = 10.39$ Å, $c = 12.17$ Å and $\alpha = 79.54°$, $\beta = 81.86°$ and $\gamma = 87.16°$

[Answer: 1149.3 Å^3]

7. Deduce the Miller indices for the plane *abc* whose intercepts on *x, y* and *z* axes are 1, ∞ and 1.

[Answer: 101]

8. Calculate the interplanar spacing of a tetragonal unit cell with unit cell parameters $a = 11.5$ Å and $c = 13.5$ Å for [111] planes.

[Answer: 10.16 Å]

9. If the edge length of the cube has a length represented by a, what is the volume of the cell in terms of a.

[Answer: a^3]

10. The distance between (110) planes in a body centred cubic structure is 0.208 nm. What is the size of the unit cell? What is the radius of the atom?

[Answer: 2.94 Å, 1.47 Å]

BONDING IN SOLIDS

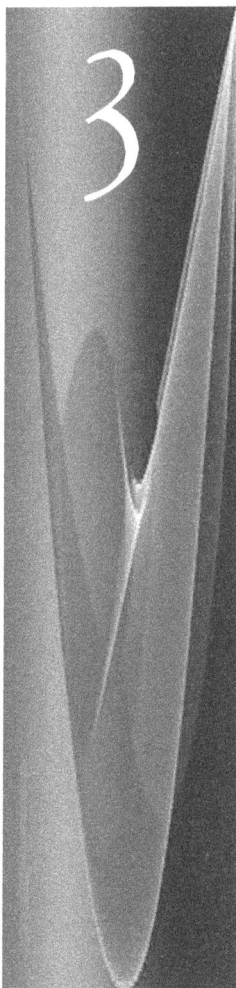

COVALENT BOND

Atoms can combine to achieve an octet of valence electrons by sharing electrons. Two fluorine atoms, for example, can form a stable F_2 molecule in which each atom has an octet of valence electrons by sharing a pair of electrons.

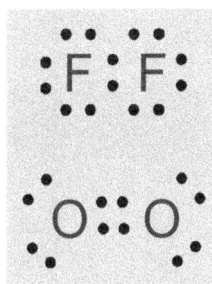

Figure 3.1 Sharing of electrons

A pair of oxygen atoms, can form an O_2 molecule in which each atom has a total of two valence electrons, by sharing two pairs of electrons (Figure 3.1). The term **covalent bond** is used to describe the bonds in compounds that result from the sharing of one or more pairs of electrons.

How Sharing of Electrons Bond Atoms?

To understand how sharing a pair of electrons can hold atoms together, let us look at the simplest covalent bond, the bond that forms when two isolated hydrogen atoms come together to form an H_2 molecule.

$$H + H \rightarrow H - H$$

An isolated hydrogen atom contains one proton and one electron held together by the force of attraction between oppositely charged particles. The magnitude of this force is proportional to the product of the charge on the electron (q_e) and the charge on the proton (q_p) divided by the square of the distance between these particles (r^2).

$$F \alpha \frac{q_e \times q_p}{r^2}$$

When a pair of isolated hydrogen atoms is brought together, two new forces of attraction appear because of the attraction between the electron on one atom and the proton on the other as shown in Figure 3.2.

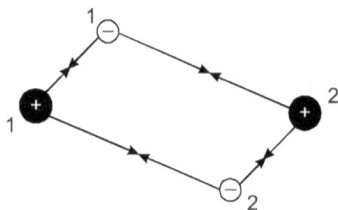

Figure 3.2   Forces of attraction

But two forces of repulsion are also created because the two negatively charged electrons repel each other, as do the two positively charged protons (Figure 3.3).

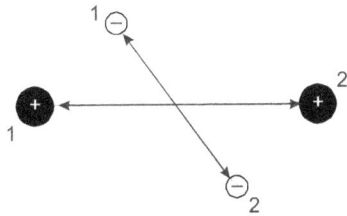

Figure 3.3 Forces of repulsion

It might seem that the two new repulsive forces would balance the two new attractive forces. If this happened, the H_2 molecule would be no more stable than a pair of isolated hydrogen atoms. But there are ways in which the forces of repulsion can be minimized. The electrons behave as if they were tops spinning on an axis. Just as there are two ways in which a top can spin, there are two possible states for the spin of an electron: $s = +1/2$ and $s = -1/2$. When electrons are paired, so that they have opposite spins, the force of repulsion between these electrons is minimized.

The force of repulsion between the protons can be minimized by the presence of the pair of electrons between the two nuclei. The distance between the electron on one atom and the nucleus of the other is now smaller than the distance between the two nuclei. As a result, the force of attraction between each electron and the nucleus of the other atom is larger than the force of repulsion between the two nuclei, as long as the nuclei are not brought too close together.

The net result of pairing the electrons when it is in between the two nuclei is a system that is more stable than a pair of isolated atoms if the nuclei are close enough together to share the pair of electrons, but not so close that repulsion between the nuclei becomes too large. The hydrogen atoms in a H_2 molecule are therefore held together (or bonded) by the sharing of a pair of electrons and this bond is the strongest when the distance between the two nuclei is about 0.074 nm. The **bond length** is the distance between the centres of the nuclei involved in the bond. The length depends on the nature of the atoms being joined (covalent radii) and **bond order** (number of shared pairs).

Table 3.1 Average covalent bond lengths in Å

C–C 1.54	C=C 1.34	C≡C 1.20
N–N 1.45	N=N 1.25	N≡N 1.10
C–H 1.10	N–H 1.01	O–H 0.96
C–O 1.43	C=O 1.20	
N–O 1.43	N=O 1.18	
C–N 1.47	C≡N 1.16	

For most covalent substances, their bond character falls between these two extremes. Bond polarity is a useful concept for describing the sharing of electrons between atoms.

- A **non-polar covalent bond** is one in which the electrons are shared equally between two atoms.

- A **polar covalent bond** is one in which *one atom has a greater attraction for the electrons than the other atom*. If this relative attraction is great enough, then the bond is an **ionic bond**.

IONIC BONDING

Ionic bonding occurs in a solid that contains both highly electronegative atoms and highly electropositive atoms. If electronegative atoms and electropositive atoms are in the same vicinity, they can both achieve noble gas configurations by transfer of valence electrons: the electropositive atoms give up some electrons, and the electronegative atoms accept them.

The tendency for charge transfer between atoms increases as the electronegativity difference (ΔEN) increases between the dissimilar atoms. After charge transfer has occurred, both atoms are ionized (they have a net charge). The electronegative atoms become negatively charged (anions), while the electropositive atoms become positively charged (cations).

When a highly electronegative atom and an electropositive atom are bonded together, an electron is transferred from the electropositive atom to the electronegative atom to form a cation and an anion respectively. The cation, being a positively charged ion, is attracted to the negatively charged anion by coulombic forces (Figure 3.4) as described by Coulomb's law:

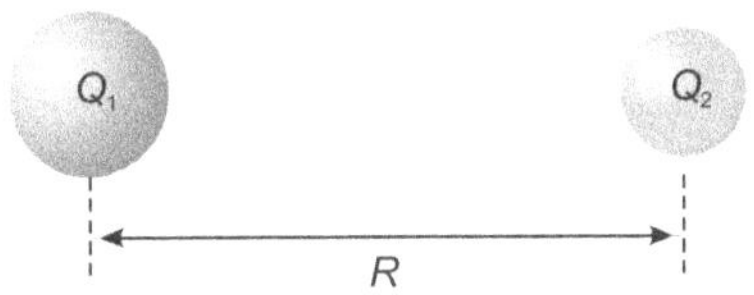

$$F \ \alpha \ \frac{Q_1 Q_2}{R^2}$$

Figure 3.4   Coulomb's law—oppositely charged species attract each other

The coulombic forces that bind cations and anions is called ionic bonding. The potential between the two charges is given by

$$E = \frac{-KQ_1 Q_2}{R}$$

A negative energy means there is an attractive interaction between the particles in the above expression. If the charges on the two ions are opposite in sign, they will attract each other. Conversely, if two charges are similar, they repel each other. Using this knowledge we can construct a graph of energy versus distance for two oppositely charged ions (Figure 3.5). At large distances, there is a negligible energy of attraction between the two ions, but as they are brought closer together, they are attracted to one another. Coulomb's law may seem to predict that the ions should be as close as possible to achieve a minimal energy state. However, the graph of energy versus distance shows that the ions are actually repelled at small distances. To explain this observation, remember that the ions' nuclei are both positively charged. When the nuclei approach each other, they repel strongly—accounting for the steep rise in potential, as the ions get closer than the bond length.

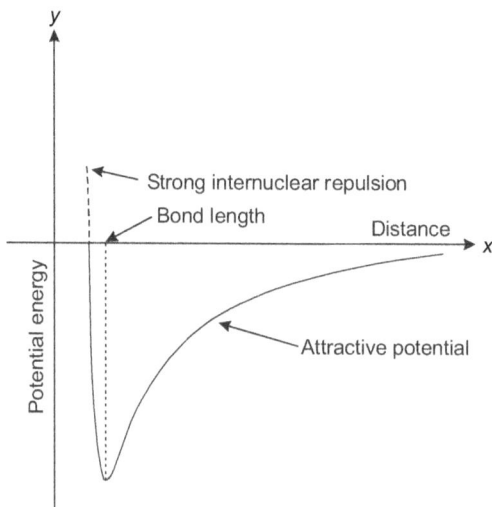

Figure 3.5   Plot of potential energy versus distance for oppositely charged ions

The depth (*y*-axis) of the minimum in the potential energy curve in Figure 3.5 represents the bond strength, and the distance (*x*-axis) at the energy minimum is the bond length. Using Coulomb's law and the bond length, one can actually predict with some accuracy the strength of an ionic bond. Performing a series of these calculations one finds that ionic compounds formed by ions with larger charges create stronger bonds and those ionic compounds with shorter bond lengths form stronger bonds.

The "strength" of an ionic bond is determined by the work required to form it. The work to form an ionic bond is composed of the following.

- the net energy required for charge transfer,
- the work done by attractive forces between cation and anion, and
- the work done against short-range repulsive forces between cation and anion.

Net Energy for Charge Transfer

$$\Delta W_{ct} = U_i$$

= (Work done to *remove* electron from shell of electropositive atom) + (Work done to *add* electron from shell of electropositive atom)

The first term is called the **ionization potential** and the second term is called the **electron affinity.** The force between two charged particles is the **coulombic force**.

$$\vec{F}_c(x) = -\frac{q_1 q_2}{4\pi \varepsilon_0 x^2} \hat{x} \tag{3.1}$$

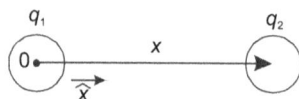

Figure 3.6   Coulombic force

The term q_i is the charge on particle i (sign and **magnitude). Example:** Na^+ has $q = 1.608 \times 10^{-19}$ C. Greek epsilon in denominator in equation 3.1 is the dielectric permittivity of free space ($\varepsilon_0 = 8.85 \times 10^{-12}$ C²/Nm²). The force is <0 (repulsive) if the particles have the same charge, but is >0 (attractive) if they have opposite charge. As two isolated ions approach each other from infinite separation, their electrons begin to become part of the same system. When the ions are close enough together that their electron orbitals overlap, the electrons become fully interacting. Interacting electrons are subjected to the Pauli Exclusion Principle, according to which no two electrons can have all the four quantum numbers to be the same. One electron must adopt a different quantum number(s) which *increase* its energy. As the electron energies increase, it becomes less and less favourable for the ions to continue approaching one another. A repulsive force arises that tends to push the ions apart. In addition to the electron repulsion, when electron clouds overlap, the nuclei begin to be repelled from each other because they have the same charge. The repulsive force is active only over very short distance (< 1 Å), but it is very strong.

Equation for the repulsive force as a function of separation between ions:

$$\vec{F}_r(x) = -\frac{K}{x^m}\hat{x} \tag{3.2}$$

where, $m > 2$, and is usually found by experiment to be about 12.

The **equilibrium separation** is that at which the **net** force on both particles is zero.

Set $F_c(x) = -F_r(x)$ and solving for x,

$$\frac{|Z_1 Z_2| q^2}{4\pi \varepsilon_0 x^2} \hat{x} = \frac{K}{x^m} \hat{x} \tag{3.3}$$

Group terms in x:

$$x^{m-2} = \frac{4K\pi \varepsilon_0}{|Z_1 Z_2| q^2} \tag{3.4}$$

Therefore,

$$x_{eq} = \left[\frac{4K\pi \varepsilon_0}{|Z_1 Z_2| q^2} \right]^{1/(m-2)} \tag{3.5}$$

Note

- The bond length decreases as the charge on the particles increases.

- The bond length increases as the value of m decreases.

- If $m = 2$, the bond length diverges to infinity (no bonding occurs).

In a system acted on by forces, we say that a body is at equilibrium if the vector sum of forces on the body is zero. Since force is the slope of a plot of energy versus distance, we can also say that equilibrium corresponds to an extremum of energy (specifically, a minimum).

Energy of coulombic attraction

$$\vec{F}_c(x) = \frac{|Z_1 Z_2| q^2}{4\pi \varepsilon_0 x^2} \hat{x} \tag{3.6}$$

Integrate over distance as one ion approaches another from infinite separation:

$$\Delta W_a(x) = U_a(x) = \int_{\infty}^{x} \vec{F}_a \cdot d\vec{x} = -\frac{|Z_1 Z_2| q^2}{4\pi \varepsilon_0 x} \tag{3.7}$$

Short-range repulsive energy

$$\vec{F}_r(x) = -\frac{K}{x^m} \hat{x} \tag{3.8}$$

Again, the energy is found by integrating as the ions approach from infinite separation to a distance x,

$$\Delta W_r(x) = U_r(x) = \int_{\infty}^{x} \vec{F}_r \cdot d\vec{x} = \frac{C}{x^{m-1}} \tag{3.9}$$

Total bond energy

$$U(x) = U_i - \frac{|Z_1 Z_2| q^2}{4\pi \varepsilon_0 x} + \frac{C}{x^{m-1}} \tag{3.10}$$

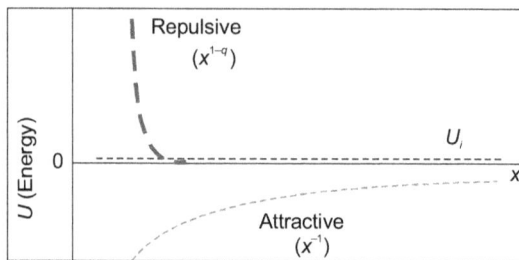

Figure 3.7 Ionic bonds: bond-potential curve

The structure of NaCl consists of two interpenetrating fcc lattices of Na^+ and Cl^- ions as shown in the Figure 2.6. Thus each Na^+ ion is surrounded by 6 Cl^- ions and vice versa. This structure suggests that there is a strong attractive

Coulomb interaction between nearest-neighbour ions, which is responsible for the ionic bonding. To calculate the binding energy we need to include Coulomb interactions with all atoms in the solid. Also we need to take into account the repulsive energy, which we assume to be exponential. Thus the interaction between two atoms i and j in a lattice is given by

$$U_{ij} = \lambda e^{-\frac{r_{ij}}{\rho}} \pm \frac{q^2}{r_{ij}} \qquad (3.11)$$

Here r_{ij} is the distance between the two atoms, q is the electric charge on the atom, the (+) sign is taken for the like charges and the (−) sign for the unlike charges. The total energy of the crystal is the sum over i and j so that

$$U = \frac{1}{2}\sum_{i,j} U_{ij} = \frac{N}{2}\sum_j U_{ij} = N\sum_j \left(\lambda e^{-r_{ij}/\rho} \pm \frac{q^2}{r_{ij}} \right) \qquad (3.12)$$

In this formula, 1/2 is due to the fact that each pair of interactions should be counted only once. The second equality results from the fact that in the NaCl structure the sum over j does not depend on whether the reference ion i is positive or negative, which gives the total number of atoms. The latter divided by two gives the number of molecules N, composed of both positive and negative ions.

We assume for simplicity that the repulsive interaction is non-zero only for the nearest neighbours (because it drops down very quickly with the distance between atoms). In this case we obtain

$$U = N \left(z\lambda e^{-R/\rho} - \alpha \cdot \frac{q^2}{R} \right) \qquad (3.13)$$

where,

R is the distance between the nearest neighbours,

z is the number of the nearest neighbours and

α is the **Madelung constant**.

$$\alpha = \sum_{j \neq i} \frac{(\pm 1)}{p_{ij}} \qquad (3.14)$$

where, p_{ij} is defined by $r_{ij} = p_{ij} R$.

The value of the Madelung constant plays an important role in the theory of ionic crystals. In general it is impossible to compute the Madelung constant analytically. A powerful method for calculating lattice sums was developed by Ewald and known as Ewald summation.

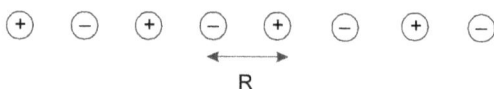

Figure 3.8 A one-dimensional lattice of ions of alternating sign

$$\alpha = 2\left[1 - \frac{1}{2} + \frac{1}{3} - \frac{1}{4} + \frac{1}{5} - \ldots\right] = 2\ln 2 \qquad (3.15)$$

The terms within the square brackets in the above equation arise from

$$\ln(1 + x) = \sum_{n=1}^{\infty} (-1)^{n-1} \frac{x^n}{n} \qquad (3.16)$$

when, $x = 1$.

In three dimensions, calculating the series is much more difficult. The values of the Madelung constant for various solids are calculated, tabulated and can be found in literature ($\alpha_{NaCl} = 1.75$). Now we can calculate the distance between the nearest neighbours for the NaCl type lattice. At the equilibrium, the derivative $dU/dR = 0$, so that

$$-\frac{z\lambda}{\rho}e^{-R_0/\rho} + \frac{\alpha q^2}{R_0^2} = 0 \qquad (3.17)$$

or

$$R_0^2 e^{-R_0/\rho} = \frac{\alpha\rho q^2}{z\lambda} \qquad (3.18)$$

This relationship determines the equilibrium separation R_0 in terms of the parameters ρ and λ of the repulsive potential. The cohesive energy per atom of the ionic solid can be written as follows:

$$U_0 = \frac{\alpha N \rho q^2}{R_0^2} - \frac{\alpha N q^2}{R_0} = -\frac{\alpha N q^2}{R_0}\left(1 - \frac{\rho}{R_0}\right) \qquad (3.19)$$

Let us estimate the magnitude of the cohesive energy in NaCl. The Madelung constant, $\alpha = 1.75$. The interatomic distance is $R_0 = a/2$ (2.81 Å). The charge $q = e$. The repulsive interaction has a very short range of the order of $\rho = 0.1R_0$. Substituting these in equation (3.19) one obtains,

$$\frac{U_0}{N} \approx -\frac{\alpha e^2}{R_0}\left(1 - \frac{0.1R_0}{R_0}\right) \approx -8\,\text{eV} \qquad (3.20)$$

(**Note**: in *SI* units $e^2 \Rightarrow e^2/4\pi\varepsilon_0$: $\varepsilon_0 = 10^7/4\pi C^2$)

We see that the typical value of the binding energy per pair of atoms is about 8 eV. This implies that ionic bond is very strong. Experimentally, this strength is characterized by the relatively high melting temperatures. For example, the melting temperature of NaCl is about 1100°C, while the melting temperature for the Na metal is about 400°C (weaker metallic bond).

SIMILARITIES AND DIFFERENCES BETWEEN IONIC AND COVALENT COMPOUNDS

There is a significant difference between the physical properties of NaCl and Cl_2, as shown in Table 3.2 below, which results from the difference between the ionic bonds in NaCl and the covalent bonds in Cl_2 are given.

Table 3.2   Some physical properties of NaCl and Cl_2

Properties	NaCl	Cl_2
Phase at room temperature	Solid	Gas
Density	2.165 g/cm³	0.003214 g/cm³
Melting point	801°C	−100.98°C
Boiling point	1413°C	−34.6°C
Ability of aqueous solution to conduct electricity	Conducts	Does not conduct

Each Na^+ ion in NaCl is surrounded by six Cl^- ions, and vice versa, as shown in Figure 3.9. Removing an ion from this compound therefore involves breaking at least six bonds. Some of these bonds would have to be broken to melt NaCl, and they would all have to be broken by some external energy to boil this compound. As a result, ionic compounds such as NaCl tend to have high melting points and boiling points. Ionic compounds are therefore solids at room temperature.

Chlorine consists of molecules in which one atom is tightly bound to another, as shown in Figure 3.9. The covalent bonds within these molecules are at least as strong as an ionic bond, but we do not have to break these covalent bonds to separate one Cl_2 molecule from another. As a result, it is much easier to melt Cl_2 to form a liquid or boil it to form a gas, and Cl_2 is a gas at room temperature.

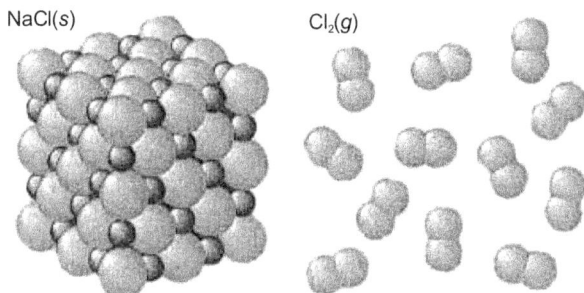

Figure 3.9 Ionic and covalent compounds

When two chlorine atoms come together to form a covalent bond, each atom contributes one electron to form a pair of electrons shared equally by the two atoms, as shown in Figure 3.10. When a sodium atom combines with a chlorine atom to form an ionic bond, each atom still contributes one electron to form a pair of electrons, but the two atoms do not share this pair of electrons. The electrons spend most of their time on the chlorine atom.

The formation of a covalent bond

$$:\overset{..}{\underset{..}{Cl}}\cdot \ + \ \cdot\overset{..}{\underset{..}{Cl}}: \ \longrightarrow \ :\overset{..}{\underset{..}{Cl}}-\overset{..}{\underset{..}{Cl}}:$$

The formation of an ionic bond

$$Na^+ \ + \ \cdot\overset{..}{\underset{..}{Cl}}: \ \longrightarrow \ [Na]^+ \ [:\overset{..}{\underset{..}{Cl}}:]^-$$

Figure 3.10 Formation of covalent and ionic bonds

Ionic and covalent bonds differ in the extent to which a pair of electrons is shared by the atoms that form the bond. When one of the atoms is much better at drawing electrons towards itself than the other, the bond is **ionic**. When the atoms are approximately equal in their ability to draw electrons towards themselves, the atoms share the pair of electrons more or less equally, and the bond is **covalent**. As a rule of the thumb, metals often react with non-metals to form ionic compounds or salts, and non-metals combine with other non-metals to form covalent compounds. This rule of thumb is useful, but it is also naive, for two reasons.

- The only way to tell whether a compound is ionic or covalent is to measure the relative ability of the atoms to draw electrons in a bond towards themselves.

- Any attempt to divide compounds into just two classes (ionic and covalent) is doomed to failure because the bonding in many compounds falls between these two extremes.

The first limitation is the basis of the concept of electronegativity. The second serves as the basis for the concept of polarity.

The relative ability of an atom to draw electrons in a bond towards itself is called the **electronegativity** of the atom. Atoms with large electronegativities (such as F and O) attract the electrons in a bond better than those that have small electronegativities (such as Na and Mg).

USING ELECTRONEGATIVITY TO IDENTIFY IONIC, COVALENT, AND POLAR COVALENT COMPOUNDS

When the difference between the electronegativities of the elements in a compound is relatively large, the compound is best classified as **ionic**.

Example: NaCl, LiF, and $SrBr_2$ are good examples of ionic compounds (Table 3.3). In each case, the electronegativity of the non-metal is at least two units larger than that of the metal.

Table 3.3 Electronegativity difference and ionic character

NaCl		LiF		SrBr$_2$	
Cl	$EN = 3.16$	F	$EN = 3.98$	Br	$EN = 2.96$
Na	$EN = 0.93$	Li	$EN = 0.98$	Sr	$EN = 0.95$
	$\Delta EN = 2.23$		$\Delta EN = 3.00$		$\Delta EN = 2.01$

We can therefore assume a net transfer of electrons from the metal to the nonmetal to form positive and negative ions and write the Lewis structures of these compounds as shown in Figure 3.11.

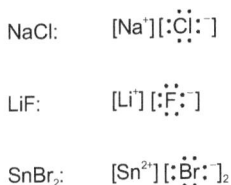

NaCl: $[Na^+][:\overset{..}{\underset{..}{Cl}}:^-]$

LiF: $[Li^+][:\overset{..}{\underset{..}{F}}:^-]$

SnBr$_2$: $[Sn^{2+}][:\overset{..}{\underset{..}{Br}}:^-]_2$

Figure 3.11 Lewis structures

All these compounds have high melting points and boiling points, as might be expected for ionic compounds (Table 3.4).

Table 3.4 Melting point (MP) and boiling point (BP) for ionic compounds

	NaCl	LiF	SrBr$_2$
MP	801°C	846°C	657°C
BP	1413°C	1717°C	2146°C

They also dissolve in water to give aqueous solutions that conduct electricity, as would be expected. When the electronegativities of the elements in a compound are about the same, the atoms share electrons, and the substance is **covalent**. Examples of covalent compounds include methane (CH_4), nitrogen dioxide (NO_2), and sulphur dioxide (SO_2) are shown in Table 3.5.

Table 3.5   Electronegativity difference and covalent character

CH_4		NO_2		SO_2	
C	*EN* = 2.55	O	*EN* = 3.44	O	*EN* = 3.44
H	*EN* = 2.20	N	*EN* = 3.04	S	*EN* = 2.58
	ΔEN = 0.35		ΔEN = 0.40		ΔEN = 0.86

These compounds have relatively low melting points and boiling points, as might be expected for covalent compounds, and they are all gases at room temperature (Table 3.6).

Table 3.6   Melting point (MP) and boiling point (BP) for covalent compounds

	CH_4	NO_2	SO_2
MP	–182.5°C	–163.6°C	–75.5°C
BP	–161.5°C	–151.8°C	–10°C

Inevitably, there must be compounds that fall between these extremes. For these compounds, the difference between the electronegativities of the elements is large enough to be significant, but not large enough to classify the compound as ionic. Consider water, for example (Table 3.7).

Table 3.7 Electronegativity difference of O and H

O	$EN = 3.44$
H	$EN = 2.20$
	$\Delta EN = 1.24$

Water is neither purely ionic nor purely covalent. It does not contain positive and negative ions, as indicated by the Lewis structure on the left in Figure 3.12. But the electrons are not shared equally, as indicated by the Lewis structure on the right in this figure. Water is best described as a **polar compound**. One end, or pole, of the molecule has a partial positive charge (δ^+), and the other end has a partial negative charge (δ^-).

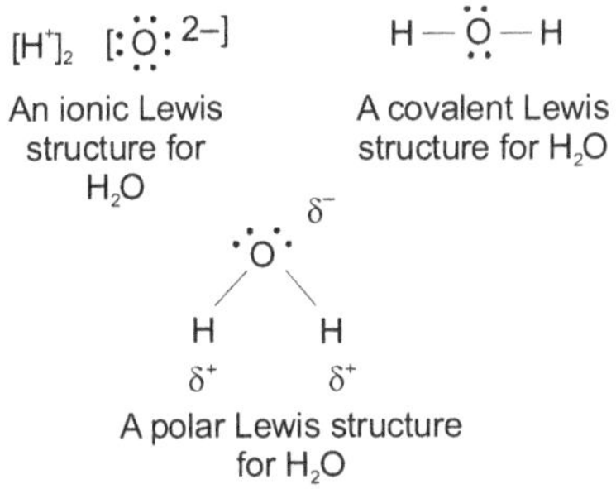

$[H^+]_2$ $[:\overset{..}{\underset{..}{O}}:^{2-}]$ $H-\overset{..}{\underset{..}{O}}-H$

An ionic Lewis   A covalent Lewis
structure for   structure for H_2O
H_2O

A polar Lewis structure
for H_2O

Figure 3.12 Water—a polar compound

As a rule, when the difference between the electronegativities of two elements is less than 1.2, we assume that the bond between atoms of these elements is **covalent**. When the difference is larger than 1.8 (Table 3.8) the bond is assumed to be **ionic**. Compounds for which the electronegativity difference is between 1.2 and 1.8 are best described as **polar** or **polar covalent**.

Table 3.8 ΔEN and bond character

Covalent	ΔEN	< 1.2	
Polar	$1.2 <$	ΔEN	< 1.8
Ionic	ΔEN	> 1.8	

TRANSITION BETWEEN COVALENT AND IONIC BONDING

Covalent and ionic bonds are end-member models. Transition states exist between them, and few if any chemical bonds are perfectly ionic or covalent. The transition between covalent and ionic bonds is monotonous and continuous. The percentage of ionic character of a bond is a function of the difference between the electronegativities of the elements in the bond, as shown in Figure 3.13.

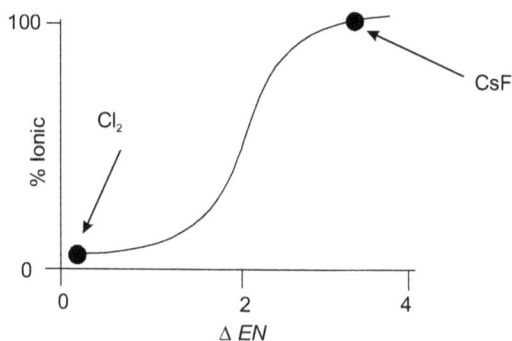

Figure 3.13   Transition continuum

METALLIC BOND

Metals tend to have high melting points and boiling points suggesting strong bonds between the atoms. Even a metal like sodium (melting point 97.8°C) melts at a considerably higher temperature than the element (Neon), which precedes it in the periodic table.

Sodium has the electronic structure $1s^2 2s^2 2p^6 3s^1$. When sodium atoms come together, the electron in the 3s atomic orbital of one sodium atom shares space with the corresponding electron on a neighbouring atom to form a molecular orbital, in much the same sort of way that a covalent bond is formed.

The difference, however, is that each sodium atom is being touched by eight other sodium atoms, and the sharing occurs between the central atom and the 3s orbitals on all of the eight other atoms. And each of these eight is in turn being touched by eight sodium atoms, which in turn are touched by eight atoms, and so on until one has taken in all the atoms in that lump of sodium.

All of the 3s orbitals on all of the atoms overlap to give a vast number of molecular orbitals which extend over the whole piece of metal. There have to be huge numbers of molecular orbitals, of course, because any orbital can only hold two electrons.

The electrons can move freely within these molecular orbitals, and so each electron becomes detached from its parent atom. The electrons are said to be **delocalized** (Figure 3.14). The metal is held together by the strong forces of attraction between the positive nuclei and the delocalized electrons.

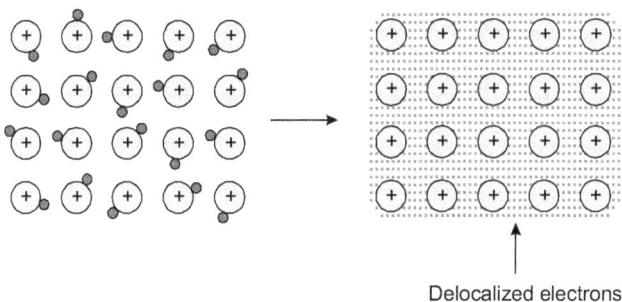

Delocalized electrons

Figure 3.14 Metallic bonding

This is sometimes described as "an array of positive ions in a sea of electrons." Each positive centre in Figure 3.14 represents the positive ions (after the delocalization of electrons). The electron has not been lost—it may no longer have an attachment to a particular atom, but it is still there in the structure. Sodium metal is therefore written as Na but not Na^+.

Metallic Bonding in Magnesium

Magnesium possesses stronger metallic bonds and so a higher melting point. Magnesium has the outer electronic structure $3s^2$. Both of these electrons become delocalized, so the "sea of electrons" has twice the electron density as it does in sodium. The remaining "ions" also have twice the charge and so there will be more attraction between "ions" and "sea of electrons".

More realistically, each magnesium atom has one more proton in the nucleus than a sodium atom has, and so not only will there be a greater number of delocalized electrons, but also a greater attraction for them.

Magnesium atoms have a slightly smaller radius than sodium atoms, and so the delocalized electrons are closer to the nuclei. Each magnesium atom also has twelve nearest neighbours rather than sodium's eight. Both of these factors increase the strength of the bond still further.

Metallic Bonding in Transition Elements

Transition metals tend to have particularly high melting points and boiling points. The reason is that they can involve the 3d electrons in the delocalization as well as the 4s. The more the electrons involved, the stronger the attractions tend to be.

Metal atoms have only a few valence electrons, which are loosely bound to the nucleus. When many atoms are brought together, these valence electrons are lost from individual atoms and are shared by all the atoms, i.e., they are **delocalized** and form an **electron gas** which fills the space between the atoms (Figure 3.15c). Attraction between the negative charge of this electron gas and the positively charged metal ions is enough to hold the structure together. The bonding is non-directional, so the metal ions try to get as close as possible together leading to **close-packed** crystal structures with high coordination numbers. The electron gas acts like "glue". The main source of the "glue" is lowering of the energy of the valence electrons in a metal as compared to the free atoms (explained based on the uncertainty principle). Because there is no directionality in the bonds, the metal ions are able to move with respect to each other, so the metals tend to be ductile. The large number of "free" electrons in the gas can easily move under the influence of an electric field; therefore, they are good conductors of electricity. The free "*s*" electrons can easily transfer energy through the crystal (good thermal conductors). Transition metals like Fe, Ni, Ti, and Co have also 3d electrons which are more localized and create covalent bonds.

The electron pairs shared between two atoms **are not necessarily shared equally**. For example in Cl_2 the shared electron pairs is shared equally. In NaCl the 3s electron is stripped from the Na atom and is incorporated into the electronic structure of the Cl atom—and the compound is most accurately described as consisting of individual Na^+ and Cl^- ions.

In summary, metallic, covalent, and ionic bonding can be picturized as shown in Figure 3.15.

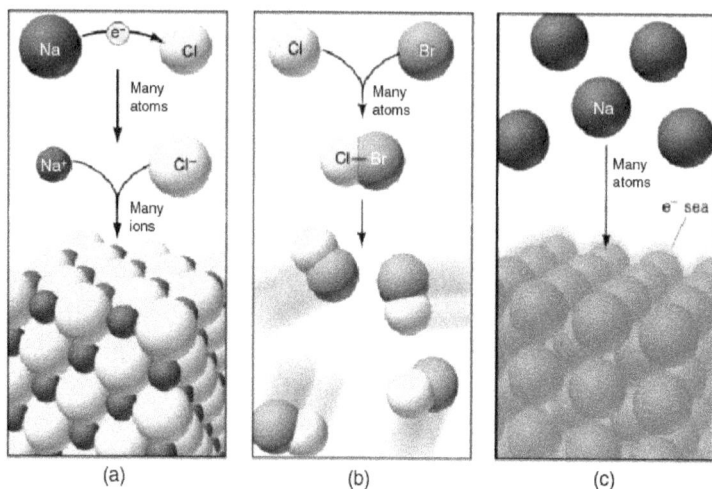

Figure 3.15 Different types of bonding. (a) Ionic bonding (b) Covalent bonding (c) Metallic bonding

Bond Order

Bond order is a measure of how many pairs of electrons are shared between two atoms to bond them together. Recall that a chemical bond is a sharing of electron pairs between two atoms. For molecules without resonance the bond order is easy. A single bond has a bond order of 1 because there is one pair of electrons being shared between the two atoms. A double bond has a bond order of 2 because the two atoms share two pairs of electrons. Likewise, in a triple bond the bond order is 3 because three pairs of electrons are shared between the two atoms bound together. Things are a little more complicated in molecules with resonance. To determine the bond order when there is resonance we must recognize that some of the electron pairs are split between more than one pair of atoms.

For example, O_3 is a resonance hybrid molecule. The bond order of each O–O is 1.5. The value 1 is from the single bond and the value 0.5 is from the pair involved in the resonance, e.g. SO_2, SO_3.

In more complicated cases we can calculate the bond order by a formula. Let n atom pairs be the number of atom pairs of a specific type involved in the resonance, and n electron pairs be the number of pairs of electrons involved in the resonance, then the bond order is n electron pairs/n atom pairs.

Bond Length

The bond length is defined as the distance between the nuclei of the two atoms involved in the bond. There are three effects that govern bond lengths:

i. Bonds with higher bond orders tend to be shorter. So, for example, a $C \equiv C$ bond is shorter than a $C=C$ bond which is shorter than a C–C bond. Likewise, a carbon–carbon bond in benzene (C_6H_6) is shorter than an unhybridized carbon–carbon bond, but longer than an unhybridized $C=C$ bond.

ii. Bond lengths tend to increase as we go down a column in the periodic table. That is, bonds between larger atoms tend to be longer. An S–H bond would be expected to be longer than an O–H bond because S is a larger atom than O. Likewise, the H–I bond is longer than the H–Br bond which is longer than the H–Cl bond which in turn is longer than the H–F bond.

iii. Bond lengths tend to decrease as we go to the right in a row in the periodic table. C–H bonds are longer than N–H bonds which are longer than O–H bonds which in turn are longer than F–H bonds.

Bond Energy

Energy is released when isolated atoms form a covalent bond. Energy must be required to break the bond apart. The **bond energy** is the amount of energy necessary to break one mole of covalent bonds into isolated gaseous species; it is sometimes called the **bond dissociation energy** or **bond enthalpy**. As the exact bond energy of a given bond varies somewhat with the species that are linked to the other side of the central atom, average bond energies are often quoted. The **enthalpy**, H, of a reaction can be estimated as the sum of the bond energies of the reactants minus the sum of the bond energies of the products.

Some electron pairs in a covalent bond are not shared equally. (The electron pairs in a bond are shared equally only when there is a bond between two atoms of the same element.) The ability of an atom to pull electrons towards itself is called **electronegativity**. The more electronegative atoms get a larger share of the electrons.

Electronegativity increases from left to right across a row on the periodic table. Electronegativity decreases as we go down a column on the periodic table. The Pauling scale is the most commonly used. Fluorine (the most electronegative element) is assigned a value of 4.0, and values range down to caesium and francium, which are the least electronegative at 0.7.

1.   *What happens if two atoms of equal electronegativity bond together?*

Consider a bond between two atoms, A and B (Figure 3.16).

$$A \underline{\quad\overset{\bullet}{\underset{\bullet}{\quad}}\quad} B$$

Figure 3.16 Bonding of atoms of equal electronegativity

If the atoms are equally electronegative, both have the same tendency to attract the bonding pair of electrons, and so it will be found on average half way between the two atoms. To get a bond like this, A and B would usually have to be the same atom. We will find this sort of bond in, for example, H_2 or Cl_2 molecules.

2. *What happens if B is slightly more electronegative than A?*

B will attract the electron pair rather more than A does.

$$\overset{\delta+}{A} \overset{\bullet}{\underset{\bullet}{\rule{3cm}{0.4pt}}} \overset{\delta-}{B}$$

Figure 3.17 Bonding of atoms of slight electronegativity difference

That means that the B end of the bond has more than its fair share of electron density and so becomes slightly negative. At the same time, the A end (rather short of electrons) becomes slightly positive. In Figure 3.17 "$\delta-$" (read as "delta") means "slightly negative", so "$\delta+$" means "slightly positive". This is described as a **polar bond**. A polar bond is a covalent bond in which there is a separation of charge between one end and the other; in other words in which one end is slightly positive and the other slightly negative. The hydrogen–chlorine bond in HCl or the hydrogen–oxygen bonds in water are typical. Polar covalent bonds form a dipole. We say that they have a dipole moment. A **dipole moment** is a charge multiplied by a distance (the distance between the separated charges). Dipole moment is a vector quantity. That is, it has a magnitude and a direction. When we add dipole moments we must include the direction information. Molecules with a dipole moment are called polar molecules.

Examples: HF, H_2O_2, N_2, CO_2, H_2O, NH_3, CH_4

3. *What happens if B is a lot more electronegative than A?*

In this case, the electron pair is dragged right over to B end of the bond. To all intents and purposes, A has lost control of its electron, and B has complete control over both electrons. Ions have been formed.

Figure 3.18   Bonding of atoms with large electronegativity differences

Thus the electronegativity differences tell us whether bonds are ionic or covalent. If the electronegativity difference is 1.9 or less, the bond is (polar) covalent. If the electronegativity difference is 2.0 or greater, the "bond" is ionic. Polar covalent means that the electrons are not shared equally.

Some examples are given in the following table.

Compounds	Electronegativity differences
NaCl	3.0 – 0.9 = 2.1
MgO	3.5 – 1.2 = 2.3
O_2	3.5 – 3.5 = 0.0
C–N	3.0 – 2.5 = 0.5
C–H	2.5 – 2.1 = 0.4

So far we have only been dealing about the polarity of individual bonds. What is of most importance to us is the polarity of a molecule. That is, is one end of a molecule more negative than the other? We cannot determine the polarity of a molecule until we learn how to determine the shapes of molecules. For example, in CCl_4, each bond is polar.

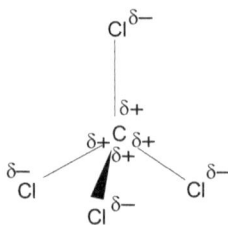

Figure 3.19   Polar bond in CCl_4

The molecule as a whole, however, is not polar, in the sense that it does not have an end (or a side), which is slightly negative, and one which is slightly positive. The whole of the outside of the molecule is somewhat negative, but there is no overall separation of charge from top to bottom, or from left to right.

DIFFERENCE BETWEEN POLAR BONDS AND POLAR MOLECULES

The difference between the electronegativities of chlorine (*EN* = 3.16) and hydrogen (*EN* = 2.20) is large enough to assume that the bond in HCl is polar.

$$\delta+ \quad \delta-$$
$$H - Cl$$

Because it contains only this one bond, the HCl molecule can also be described as polar. The polarity of a molecule can be determined by measuring a quantity known as the **dipole moment** (μ), which depends on two factors: 1) the magnitude of the separation of charge and 2) the distance between the negative and positive poles of the molecule. Dipole moments are reported as units of **debye** (d). The dipole moment for HCl is small = 1.08 d. This can be understood by noting that the separation of charge in the HCl bond is relatively small (ΔEN = 0.96) and that the H–Cl bond is relatively short.

C–Cl bonds (ΔEN = 0.61) are not as polar as H–Cl bonds (ΔEN = 0.96), but they are significantly longer. As a result, the dipole moment for CH_3Cl is about the same as HCl = 1.01 d. At first glance, we might expect a similar dipole moment for carbon tetrachloride (CCl_4), which contains four polar C–Cl bonds. The dipole moment of CCl_4, however, is 0. This can be understood by considering the structure of CCl_4 shown in Figure 3.20. The individual C–Cl bonds in this molecule are polar, but the four C–Cl dipoles cancel each other. Carbon tetrachloride therefore illustrates an important point: Not all molecules that contain polar bonds have a dipole moment.

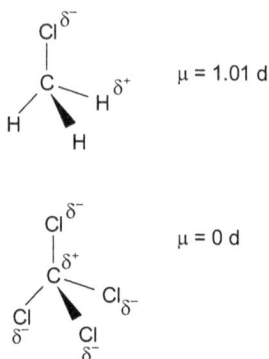

$\mu = 1.01$ d

$\mu = 0$ d

Figure 3.20   Dipole moment of polar bonds

The covalent, ionic, and metallic bonds are all quite strong, with binding energies of the order of the Coulomb energy of two electrons a few angstroms apart, a few eV/atom. There are also some much weaker forces between atoms that are responsible for liquefaction and crystallization of mutual molecules or rare gas atoms. The most important for our biological molecules is the **van der Waals force**, so called because it is responsible for the deviations from ideal gas behaviour studied by van der Waals. The force arises because even a spherically symmetrical atom or molecule has a fluctuating electric dipole moment due to electronic zero-point motion. This can induce a dipole moment in the neighbouring molecules because they are coupled via the electromagnetic field, and this correlation leads to an average net attraction. Like the metallic bond, this force is undirected and not saturable. The rare gas solids are therefore, like metals, largely determined by the packing of spheres; and organic molecules crystallize in such a way as to pack together their shapes most compactly.

The weaker bonds are essentially electrostatic, forces like van der Waals forces or hydrogen forces or hydrogen bond. van der Waals forces are present in all compounds. Together with hydrogen bonds they play a primary role in molecular crystal structure. **Hydrogen bonds** are fundamentals for the cohesion of many compounds like proteins, molecular crystals, hydrated salts, etc.

van der Waals bond, a non-specific attractive force, comes into play when any two atoms are 3 to 4 Å apart. Though weaker and less specific than electrostatic and hydrogen bonds, van der Waals bonds are no less important in biological systems. The basis of a van der Waals bond is that the distribution of electronic charge around an atom changes with time. At any instant, the charge distribution is not perfectly symmetric. This transient asymmetry in the electronic charge around an atom encourages a similar asymmetry in the electronic distribution around its neighbouring atoms. The resulting attraction between a pair of atoms increases as they come closer, until they are separated by the **van der Waals contact distance**.

At a shorter distance, very strong repulsive forces become dominant because the outer electron clouds overlap. The contact distance between an oxygen and carbon atom, for example, is 3.4 Å, which is obtained by adding 1.4 and 2.0 Å, the contact radii of the O and C atoms.

The van der Waals bond energy of a pair of atoms is about 1 kcal/mol. It is considerably weaker than a hydrogen/ electrostatic bond, which is in the range of 3 to 7 kcal/mol. A single van der Waals bond counts for very little because its strength is only a little more than the average thermal energy of molecules at room temperature (0.6 kcal/mol). Furthermore, the van der Waals force fades rapidly when the distance between a pair of atoms becomes even 1 Å greater than their contact distance. It becomes significant only when numerous atoms in one of a pair of molecules can simultaneously come close to many atoms of the other. This can happen only if the shape of the molecules matches. In other words, effective van der Waals interaction depend on steric complementarity. Though there is virtually no specificity in a single van der Waals interaction, specificity arises when there is an opportunity to make a large number of van der Waals bonds simultaneously. Repulsions between atoms closer than van der Waals contact distance are as important as attractions for establishing specificity.

WHAT ARE INTERMOLECULAR ATTRACTIONS?

Intermolecular attractions are attractions between one molecule and a neighbouring molecule. The forces of attraction which hold an individual molecule together (for example, the covalent bonds) are known as **intramolecular** attractions. All molecules experience intermolecular attractions, although in some cases these attractions are very weak. Even in a gas like hydrogen, H_2, if you slow the molecules down by cooling the gas, the attractions are large enough for the molecules to stick together eventually to form a liquid and then a solid.

In the case of hydrogen, the attractions are so weak that the molecules have to be cooled to 21 K (−252°C) before the attractions are enough to condense the hydrogen as a liquid. Helium's intermolecular attractions are even weaker—the molecules won't stick together to form a liquid until the temperature drops to 4 K (−269°C).

VAN DER WAALS FORCES: DISPERSION FORCES

Dispersion forces (one of the two types of van der Waals force) are also known as "London forces" (named after Fritz London who first suggested how they might arise).

ORIGIN OF VAN DER WAALS DISPERSION FORCES

Temporary Fluctuating Dipoles

Attractions are electrical in nature. In a symmetrical molecule like hydrogen, however, there does not seem to be any electrical distortion to produce positive or negative parts. But that is only true on average.

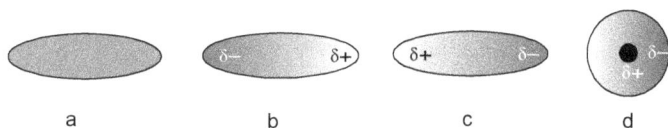

Figure 3.21   Origin of dispersion forces

The lozenge-shape of Figure 3.21a represents a small symmetrical molecule, H_2, perhaps, or Br_2. The even shading shows that on average there is no electrical distortion.

But the electrons are mobile, and at any one instant they might find themselves towards one end of the molecule, making that end $\delta-$. The other end will be temporarily short of electrons and so become $\delta+$ (Figure 3.21b).

An instant later the electrons may well have moved up to the other end, reversing the polarity of the molecule (Figure 3.21c).

This constant "sloshing around" of the electrons in the molecule causes rapidly fluctuating dipoles even in the most symmetrical molecule. It even happens in monatomic molecules—molecules of noble gases, like helium, which consist of a single atom. If both the helium electrons happen to be on one side of the atom at the same time, the nucleus is no longer properly covered by electrons for that instant (Figure 3.21d).

How do temporary dipoles give rise to intermolecular attractions?   Imagine a molecule, which has a temporary polarity being approached by one, which happens to be entirely non-polar just at that moment as shown in Figure 3.22. (In reality, one of the molecules is likely to have a greater polarity than the other at that time, and so will be the dominant one.)

Figure 3.22   Intermolecular attraction

As the right hand molecule approaches, its electrons will tend to be attracted by the slightly positive end of the left hand one.

This sets up an **induced dipole** in the approaching molecule, which is oriented in such a way that the $\delta+$ end of one is attracted to the $\delta-$ end of the other (Figure 3.23).

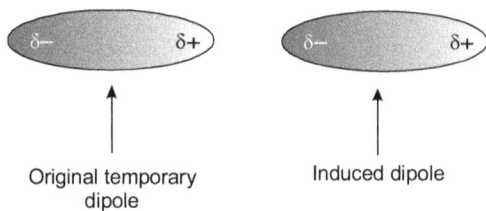

Original temporary
dipole

Induced dipole

Figure 3.23   Dipoles

An instant later, the electrons in the left hand molecule may well have moved up the other end. In doing so, they will repel the electrons in the right hand one (Figure 3.24).

Figure 3.24   Dipole–Dipole attraction

The polarity of both molecules reverses, but you still have $\delta+$ attracting $\delta-$. As long as the molecules stay close to each other, the polarities will continue to fluctuate in synchronization so that the attraction is always maintained.

There is no reason why this has to be restricted to two molecules. As long as the molecules are close together, this synchronized movement of the electrons can occur over huge numbers of molecules.

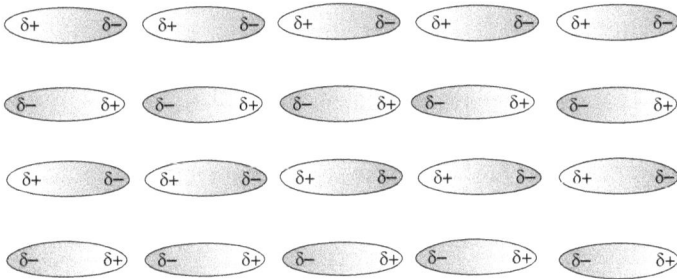

Figure 3.25   van der Waals dispersion forces in a lattice of molecules

Figure 3.25 shows how a whole lattice of molecules could be held together in a solid using van der Waals dispersion forces.

Strength of Dispersion Forces

Dispersion forces between molecules are much weaker than the covalent bonds within molecules. It is not possible to give any exact value, because the size of the attraction varies considerably with the size of the molecule and its shape. The molecular size and molecular shape affect the strength of the dispersion forces and we shall see in detail.

The boiling points of the noble gases are as follows:

Helium	−269°C
Neon	−246°C
Argon	−186°C
Krypton	−152°C
Xenon	−108°C
Radon	−62°C

All of these elements have monoatomic molecules.

The reason that the boiling points increase as we go down the group is that the number of electrons increases, so also does the radius of the atom. The more electrons we have, and the more distance over which they can move, the bigger the possible temporary dipoles and therefore the bigger the dispersion forces.

Neon   Xenon

Figure 3.26   Strength of dispersion forces

Because of the greater temporary dipoles, xenon molecules are "stickier" than neon molecules (Figure 3.26). Neon molecules will break away from each other at much lower temperatures than xenon molecules. Hence neon has the lower boiling point. This is the reason why bigger molecules have higher boiling points than smaller ones. Bigger molecules have more electrons and more distance over which temporary dipoles can develop and so the bigger molecules are "stickier".

How Molecular Shape Affects the Strength of Dispersion Forces?

The shapes of the molecules also matter. Long thin molecules can develop bigger temporary dipoles due to electron movement than short fat ones containing the same numbers of electrons.

Long thin molecules can also lie closer together. These attractions are at their most effective if the molecules are really close.

For example, the hydrocarbon molecules, butane and 2-methylpropane both have a molecular formula C_4H_{10}, but the atoms are arranged differently. In butane, the carbon atoms are arranged in a single chain, but 2-methylpropane is a shorter chain with a branch.

Butane	CH_3—CH_2—CH_2—CH_3	B.P.	−0.5°C

2-methylpropane $\quad CH_3$—CH—$CH_3 \qquad$ B.P. $\quad$ −11.7°C
$$\quad\qquad\qquad\qquad\qquad | $$
$$\qquad\qquad\qquad\qquad CH_3$$

Butane has a higher boiling point because the dispersion forces are greater. The molecules are longer (and so set up bigger temporary dipoles) and can lie closer together than the shorter, fatter 2-methylpropane molecules.

van der Waals Forces: Dipole–Dipole Interactions

A molecule like HCl has a permanent dipole (Figure 3.27) because chlorine is more electronegative than hydrogen. These permanent in-built dipoles will cause the molecules to attract each other more than they would if they had to rely only on dispersion forces.

It is important to realize that all molecules experience dispersion forces. Dipole–dipole interactions (Figure 3.27) are not an alternative to dispersion forces—they occur in addition to them. Molecules which have permanent dipoles will therefore have boiling points rather higher than molecules which only have temporary fluctuating dipoles.

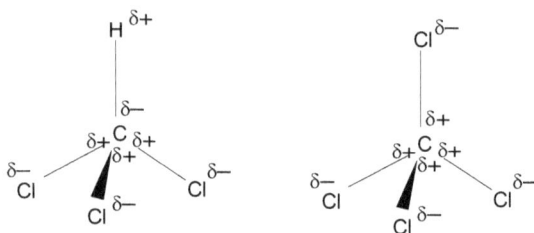

Figure 3.27   Dipole–dipole interactions

Surprisingly dipole–dipole attractions are fairly minor compared with dispersion forces, and their effect can only really be seen if we compare two molecules with the same number of electrons and the same size. For example, the boiling points of ethane (C_2H_6), and fluoromethane (CH_3F) are given in Figure 3.28.

The reason for choosing these two molecules is that both have identical number of electrons. That means that the dispersion forces in both molecules should be the same.

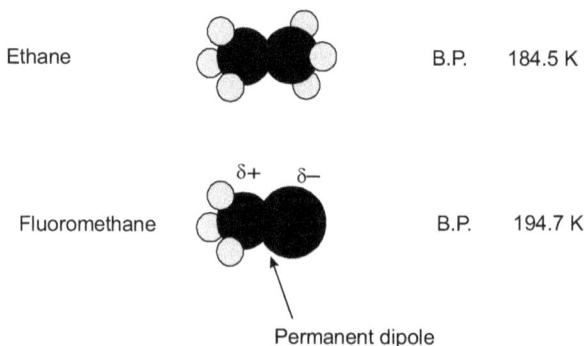

Ethane   B.P.   184.5 K

Fluoromethane   B.P.   194.7 K

Permanent dipole

Figure 3.28   Dipole and boiling point nature

The higher boiling point of fluoromethane is due to the large permanent dipole on the molecule because of the high electronegativity of fluorine. However, even given the large permanent polarity of the molecule, the boiling point has only been increased by some 10°C.

Here is another example showing the dominance of the dispersion forces. Trichloromethane, $CHCl_3$, is a highly polar molecule because of the electronegativity of the three chlorines. There will be quite strong dipole–dipole attractions between one molecule and its neighbours.

On the other hand, tetrachloromethane, CCl_4, is non-polar. The outside of the molecule is uniformly $\delta-$ in all directions. CCl_4 has to rely only on dispersion forces.

So, which has the highest boiling point? CCl_4 does, because it is a bigger molecule with more electrons. The increase in the dispersion forces more than compensates for the loss of dipole–dipole interactions.

The boiling points are

$CHCl_3$ 61.2°C

CCl_4 76.8°C

HYDROGEN BOND

The strong electrostatic attraction that occurs between molecules in which hydrogen is in a covalent bond with a highly electronegative element (i.e., fluorine, oxygen, nitrogen, chlorine) is known as **hydrogen bond**.

In covalent bonds, the electrons are shared, so that each atom gets a filled shell. When the distribution of electrons in molecules is considered in detail, it becomes apparent that the "sharing" is not always perfectly "fair": often, one of the atoms gets "more" of the shared electrons than the other does.

This occurs, in particular, when atoms such as nitrogen, fluorine, or oxygen bond to hydrogen. For example, in HF (hydrogen fluoride), the structure can be described by the "sharing" picture (Figure 3.29).

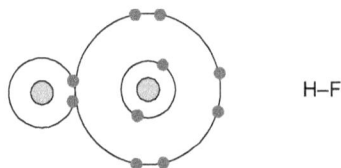

Figure 3.29   Sharing of electrons in HF

However, this structure does not tell the whole truth about the distribution of electrons in HF. Indeed, the following

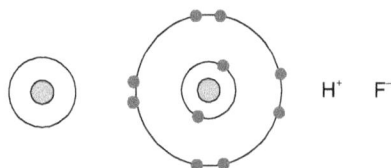

Figure 3.30   Ionic structure of HF

"ionic" structure also (Figure 3.30) respects the filled (or empty) shell rule.

In reality, HF is described by both these structures, so that the H–F bond is polar, with each atom bearing a small positive ($\delta+$) or negative ($\delta-$) charge. When two hydrogen fluoride molecules come close to each other, the unlike charges attract each other, and one gets a "molecule" of dihydrogen fluoride as shown on Figure 3.31.

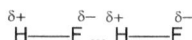

$$\overset{\delta+}{H}\!\!-\!\!-\!\!-\!\!\overset{\delta-}{F} \ldots \overset{\delta+}{H}\!\!-\!\!-\!\!-\!\!\overset{\delta-}{F}$$

Figure 3.31   Hydrogen bonding in HF

The weak "bond" between the F atom and the H is called a hydrogen bond, and is shown here as the dotted line.

In chemistry, a hydrogen bond is a type of attractive intermolecular force that exists between two partial electric charges of opposite polarity. Although stronger than most other intermolecular forces, the typical hydrogen bond is much weaker than both the ionic bond and the covalent bond. Within macromolecules such as proteins and nucleic acids, it can exist between two parts of the same molecule, and figures as an important constraint on such molecules' overall shape.

As the name "hydrogen bond" implies, one part of the bond involves a hydrogen atom. The hydrogen atom must be attached to one of the elements oxygen, nitrogen or fluorine, all of which are strongly electronegative heteroatoms. These bonding elements are known as the **hydrogen-bond donor**. This electronegative element attracts the electron cloud from around the hydrogen nucleus and, by decentralizing the cloud, leaves the atom with a positive partial charge. Because of the small size of hydrogen relative to other atoms and molecules, the resulting charge, though only partial, nevertheless represents a large charge density. A hydrogen bond results when this strong positive charge density attracts a lone pair of electrons on another heteroatom, which becomes the **hydrogen-bond acceptor**.

The hydrogen bond is not like a simple attraction between point charges. However, it possesses some degree of orientational preference, and can be shown to have some of the characteristics of a covalent bond. This covalency tends to be more extreme when acceptors bind hydrogens from more electronegative donors.

Strong covalency in a hydrogen bond raises the questions: "To which molecule or atom does the hydrogen nucleus belong?" and "Which should be labelled "donor" and which "acceptor"?" According to chemical convention, the **donor** generally is that atom to which, on separation of donor and acceptor, the retention of the hydrogen nucleus (or proton) would cause no increase in the atom's positive charge. The **acceptor** meanwhile is the atom or molecule that would become more positive by retaining the positively charged proton. Liquids that display hydrogen bonding are called associated liquids.

Hydrogen bonds can vary in strength from very weak (1–2 kJ mol^{-1}) to extremely strong (40 kJ mol^{-1}), so strong as to be indistinguishable from a covalent bond, as in the ion HF_2^-. Typical values include:

$$O\!-\!H... N \ (7 \ kcal/mol)$$

$$O\!-\!H...O \ (5 \ kcal/mol)$$

$$N\!-\!H... N \ (3 \ kcal/mol)$$

$$N\!-\!H...O \ (2 \ kcal/mol)$$

The length of hydrogen bonds depends on bond strength, temperature, and pressure. The bond strength itself is dependent on temperature, pressure, bond angle, and environment (usually characterized by local dielectric constant). The typical length of a hydrogen bond in water is 1.97 Å (197 pm).

Symmetric Hydrogen Bond

Symmetric hydrogen bonds have been observed recently spectroscopically in formic acid at high pressure (>GPa). Each hydrogen atom forms a partial covalent bond with two atoms rather than one. Symmetric hydrogen bonds have been postulated in ice at high pressure.

Dihydrogen Bond

The hydrogen bond can be compared with the closely related dihydrogen bond, which is also an intermolecular bonding interaction involving hydrogen atom. These structures have been known for some time, and well-characterized by crystallography; however, an understanding of their relationship to the conventional hydrogen bond, ionic bond, and covalent bond remains unclear. Generally, the hydrogen bond is characterized by a proton acceptor that is a lone pair of electrons in non-metallic atoms (most notably in the nitrogen, and chalcogen groups). In some cases, these proton acceptors may be pi-bonds or metal complexes. In the dihydrogen bond, however, a metal hydride serves as a proton acceptor; thus forming a hydrogen–hydrogen interaction. Neutron diffraction has shown that the molecular geometry of these complexes are similar to hydrogen bonds, in that the bond length is very adaptable to the metal complex/hydrogen donor system.

Advanced Theory of the Hydrogen Bond

The hydrogen bond remains a fairly mysterious object in the theoretical study of quantum chemistry and physics. Generally, the hydrogen bond can be viewed as a metric dependent electrostatic scalar field between two or more intermolecular bonds. This is slightly different from the intramolecular bound states of, for example, covalent or ionic bonds; however, hydrogen bonding is generally still a bound-state phenomenon, since the interaction energy has a net negative sum. The initial theory of hydrogen bonding proposed by Linus Pauling suggested that the hydrogen bonds had a partial covalent nature. This remained a controversial conclusion until the late 1990s when NMR techniques were employed to transfer information between hydrogen-bonded nuclei, that would only be possible if the hydrogen bond contains some covalent character. While a lot of experimental data has been recovered for hydrogen bonds in water, for example, that provide good resolution on the scale of intermolecular distances and molecular thermodynamics, the kinetic and dynamic properties of the hydrogen bond in dynamic systems remains largely mysterious.

Introduction to Hydrogen Bonding in Water

In a water molecule (H_2O), the oxygen nucleus with +8 charge attracts electrons better than the hydrogen nucleus with its +1 charge. Hence, the oxygen atom is partially negatively charged and the hydrogen atom is partially positively charged. The hydrogen atoms are not only covalently attached to their oxygen atoms but also attracted towards other nearby oxygen atoms. This attraction is the basis of the "hydrogen" bonds.

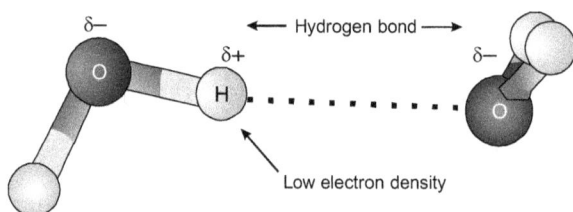

Figure 3.32   Hydrogen bond in water

The hydrogen bond in water (Figure 3.32) is a weak bond, never stronger than about a twentieth of the strength of the O–H covalent bond. It is strong enough, however, to be maintained during thermal fluctuations at, and below, ambient temperatures. The attraction of the O–H bonding electrons towards the oxygen atom leaves a deficiency on the far side of the hydrogen atom relative to the oxygen atom. The result is that the attractive force between the O–H hydrogen and the O-atom of a nearby water molecule is strongest when the three atoms are in a straight line (i.e., O–H...O) and when the O-atoms are separated by about 0.28 nm.

Each water molecule can form two hydrogen bonds involving their hydrogen atoms plus two further hydrogen bonds utilizing the hydrogen atoms attached to neighbouring water molecules. These four hydrogen bonds optimally arrange themselves tetrahedrally around each water molecule as found in ordinary ice. In liquid water, thermal energy causes bends and stretches and sometimes breaks these hydrogen bonds. However, the "average" structure of a water molecule is similar to this tetrahedral arrangement. Figure 3.33 shows such a typical "average" cluster of five water molecules. In ice this tetrahedral clustering is extensive, producing its crystalline form. In liquid water, the tetrahedral clustering is only locally found and reduces with increasing temperature. However, hydrogen bonded chains still connect liquid water molecules separated by large distances.

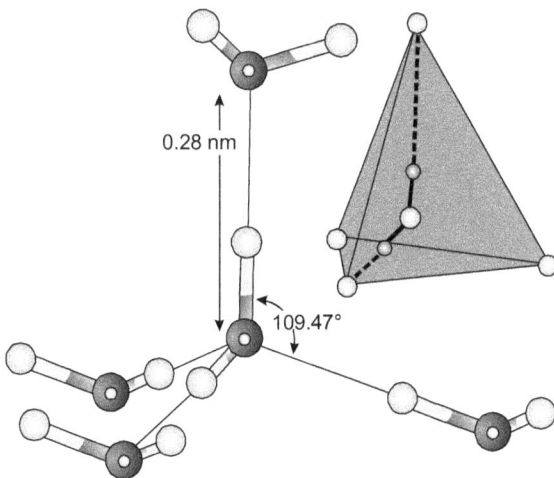

Figure 3.33   Weak O–H...O bond

There is a balance between the strength of the hydrogen bonds and the linearity that strong hydrogen bonds impose on the local structure. The stronger the bonds, the more ordered and static is the resultant structure. The structure disorder is proportional to the temperature, being smaller at lower temperatures. This is why the structure of liquid water is more ordered at low temperatures. This increase in orderliness in water as the temperature is lowered is far greater than in other liquids, due to the strength and preferred direction of the hydrogen bonds, and is the primary reason for water's rather unusual properties.

Hydrogen Bond in Biomolecules

The hydrogen bond is one of the least well-understood components in the energy decomposition that is used to predict the folding of biological complexes such as proteins. Its importance stems from its directionality and modest bonding energies midway between strong covalent and weak van der Waals bonds. For this reason, the intermolecular interaction is difficult to characterize.

As the chain of amino acids is formed, various interactions begin to take place among the amino acids along the chain. Linus Pauling and Robert Corey discovered that hydrogen bonds could form between the slightly positive amino hydrogen of one amino acid and the slightly negative carboxyl oxygen of another amino acid. Due to these H bonds, two possible structures result: an **alpha** helix and a **beta pleated sheet**. The regular, repeated configurations caused by hydrogen bonding between atoms of the polypeptide backbone is called the **secondary structure**.

In spite of substantial progress, the problem of the exact nature of the hydrogen bond remained unsolved. Pauling and others considered the hydrogen bond to be essentially an electrostatic interaction. It was not until later that Pauling proposed the quantum character of this interaction. Recently, X-ray Compton scattering measurements in ice revealed subtle oscillations in their anisotropy, which reflects quantum mechanical aspects of the hydrogen bond. In a simple quantum-mechanical descriptions of the hydrogen bond, the highest occupied molecular orbital (HOMO) of a proton acceptor fragment is as a lone pair p-orbital of the oxygen, while the lowest unoccupied molecular orbital (LUMO) of the donor neighbour molecule is an antibonding orbital. Therefore, the problem can be reduced to an effective 2 by 2 eigenvalue calculation where the HOMO–LUMO mixing can be expressed by a mixing angle.

Thomson scattering, which forms the basis of standard X-ray crystallography, can be conceptualized in essentially classical terms: an incoming electromagnetic wave induces point particles to vibrate, so that these in turn become sources of radiation with wavelength identical to that of the incident ray. These waves can then interfere and produce diffraction-like effects in a familiar manner. Compton scattering, in contrast, is a fundamentally quantum-mechanical phenomenon. The scattering event is the collision of a photon with another particle, such as an electron. The scattered photon loses energy and momentum to the particle, and therefore experiences a change in wavelength. For an electron, the set of allowed momentum states corresponds to its probability density given by the square modulus of the wave function. The more prominent a given momentum state is in a wavefunction, greater is the number of photons scattering with that momentum shift. Such a pattern of intensity versus scattering angle constitutes the Compton profile of the system. The Compton scattering is therefore a useful tool for the investigation of the quantum nature of hydrogen bond in ice and probably in other compounds such as urea, RNA, DNA and proteins. In these macromolecules, bonding between parts of the same macromolecule causes it to fold into a specific shape, which helps to determine the molecule's physiological or biochemical role. The **double helical structure** of DNA, for example, is due largely to hydrogen bonding between the base pairs, which link one complementary strand to the other and enable replication.

In proteins, hydrogen bonds form between the backbone oxygens and amide hydrogens. When the spacing of the amino acid residues participating in a hydrogen bond occurs regularly between positions i and i + 4, an alpha helix is formed. When the spacing is less, between positions i and i + 3, then a 3_{10} helix is formed. When two strands are joined by hydrogen bonds involving alternating residues on each participating strand, a **beta sheet** is formed. Hydrogen bonds also play a part in forming the tertiary structure of protein through interaction of R-groups.

Special cases of intramolecular hydrogen bonds within proteins poorly shielded from water attack and hence promoting their own dehydration, are called dehydrons.

In order to discuss in more detail the nature of the hydrogen bonding present in molecular crystals, it is necessary to be able to characterize the common structural features found. A general hydrogen bond is comprised of a donor group X–H and an acceptor A, and is referred to as X–H...A. The hydrogen bond is a long-range interaction, and thus the possibility exists for a donor group to be bonded to more than one acceptor at a time. This is called a **bifurcated bond**. In Figures 3.34 and 3.35, an example of a bifurcated donor and a bifurcated acceptor are shown. A further possibility is that of a **chelated bond**; examples of chelated bonds are shown in Figure 3.36. In both of these diagrams (Figures 3.34 and 3.35), solid lines denote chemical bonds, whilst dashed lines denote a hydrogen bond.

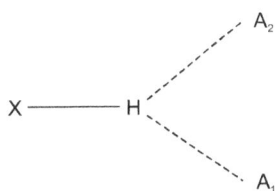

Figure 3.34 Bifurcated hydrogen bond: an example of a bifurcated donor

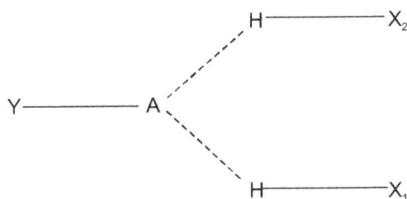

Figure 3.35 Bifurcated hydrogen bond: an example of a bifurcated acceptor

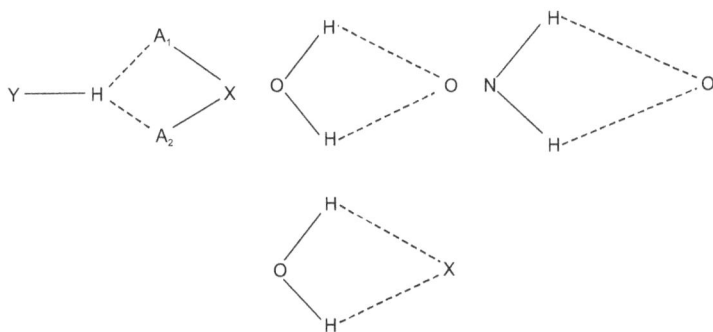

Figure 3.36 Examples of chelated bonds

Bifurcated acceptors and donors may arise in systems in which there is an excess of donors (or acceptors) relative to the number of acceptors (or donors) present. In systems in which weak hydrogen bonds are capable of forming, this leads to bifurcated donors being common.

Bifurcated hydrogen bonds (where both hydrogen atoms from one water molecule are hydrogen-bonded to some other water molecule, or one hydrogen atom simultaneously forms hydrogen bonds to two other water molecules) have just under half the strength of a normal hydrogen bond (per half the bifurcated bond) and present a low-energy route for hydrogen-bonding rearrangements. They allow the constant randomization of the hydrogen bonding within the network. However, it should be noted that they require the breakage of two hydrogen bonds; one hydrogen bond to form the bifurcated arrangement and another to make way for a different hydrogen bond to form. Any necessary rotation may also involve bending or stretching other hydrogen bonds. Bifurcation of hydrogen bonds cannot cause their net breakage and only occurs when a broken hydrogen bond releases a lone pair to accept the incoming hydrogen bond donor. Trifurcated hydrogen bonds (where one hydrogen atom simultaneously forms hydrogen bond with three other water molecules, forming a tetrahedral face) may also form but only have about one-sixth the strength of a normal hydrogen bond and require free lone pairs on all three bound water molecules and the rest of the local cluster must also be poorly hydrogen-bonded.

REVIEW QUESTIONS

1. Explain polar and non-polar covalent bonds.

2. What do you mean by electronegativity?

3. How do you identify the ionic, covalent, and polar covalent bonds using electronegativity values?

4. Explain metallic bonding.

5. What is bond order?

6. Explain the origin of van der Waals forces.

7. What are strong and weak hydrogen bonds?

8. What are the types of non-bonded interactions in biomolecules?

9. Explain bifurcated and trifurcated hydrogen bonds.

10. Explain the hydrogen bonding in water.

CRYSTAL STRUCTURE DETERMINATION—X-RAY DIFFRACTION AND OTHER DIFFRACTION TECHNIQUES

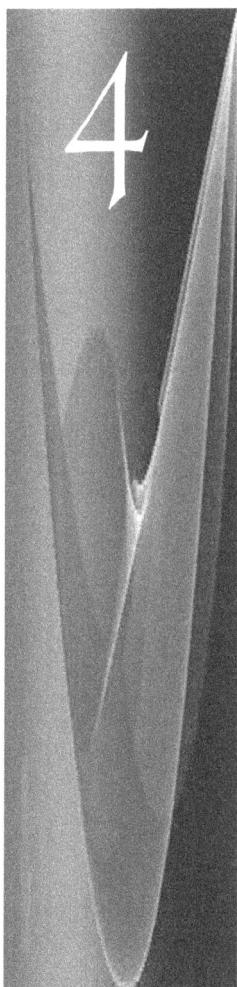

RECIPROCAL LATTICE

Many physical phenomena involve a variation with the reciprocal of distance (i.e., $1/x$). It is therefore often convenient to use a reciprocal space plot, in which one or more of the axes involve reciprocal units. A reciprocal lattice is a construction drawn in reciprocal space. A dimension of $1/d$, where d is the spacing between crystal planes, is frequently used. The most common applications are those involving diffraction ($2d \sin\theta = n\lambda$). In order to make it easier to understand the information in a diffraction pattern, we use reciprocal lattices.

Definition

The reciprocal space lattice is a set of imaginary points constructed in such a way that the direction of a vector from one point to another coincides with the direction of a normal to the space planes and the separation of those points (absolute value of the vector) is equal to the reciprocal of the real interplanar distance.

Reciprocal space is also called Fourier space, K-space, or momentum space because momentum is directly proportional to wave vector and is inversely proportional to the wavelength. Therefore each point of the reciprocal lattice represents the direction and the magnitude of momentum. The reciprocal lattice is itself a Bravais lattice, and the reciprocal of the reciprocal lattice is the original lattice.

The reciprocal lattice is therefore an essential concept for the study of crystal lattices and their diffraction properties. This concept and the relation of the direct and reciprocal lattices through the Fourier transform were first introduced in crystallography by **P.P. Ewald** (1921).

The Reciprocal Lattice Vectors

For a three-dimensional lattice, defined by its primitive vectors (a_1, a_2, a_3), its reciprocal lattice can be determined by generating its three reciprocal primitive vectors, through the formula,

$$a_1^* = 2\pi \frac{a_2 \times a_3}{a_1 \cdot (a_2 \times a_3)}$$

$$a_2^* = 2\pi \frac{a_3 \times a_1}{a_2 \cdot (a_3 \times a_1)} \qquad (4.1)$$

$$a_3^* = 2\pi \frac{a_1 \times a_2}{a_3 \cdot (a_1 \times a_2)}$$

Using column vector representation of (reciprocal) primitive vectors, the above formula can be rewritten using matrix inversion

$$[a_1^* a_2^* a_3^*]^T = 2\pi [a_1 a_2 a_3]^{-1} \qquad (4.2)$$

The factor of 2π comes naturally from the study of periodic structures. It is convenient to let the reciprocal lattice vector be 2π times the reciprocal of the interplanar distance.

This convention converts the units from periods per unit length to radians per unit length. Radians per centimetre (cm^{-1}) are widely used units. It simplifies the comparison of different periodic phenomena. In crystallography, the equivalent definition for the reciprocal lattice of a Bravais lattice is the set of all vectors K such that

$$e^{iK \cdot R} = 1$$

where, R is the position vector of all lattice points. According to this definition, the reciprocal lattice vectors become

$$a_1^* = \frac{a_2 \times a_3}{a_1 \cdot (a_2 \times a_3)} \qquad (4.3)$$

and so on for the other vectors. *Each point (hkl) in the reciprocal lattice corresponds to a set of lattice planes (hkl) in the real space lattice.* The direction of the reciprocal lattice vector corresponds to the normal of the real space planes, and the magnitude of the reciprocal lattice vector is equal to the reciprocal of the interplanar spacing of the real space planes.

Role of Reciprocal Lattice

The reciprocal lattice plays a fundamental role in most analytic studies of periodic structures, particularly in the theory of diffraction. For Bragg reflections in neutron and X-ray diffraction, the momentum difference between the incoming and diffracted X-rays of a crystal is a reciprocal lattice vector. The diffraction pattern of a crystal can be used to determine the reciprocal vectors of the lattice. Using this process, one can infer the atomic arrangement of a crystal. The **Brillouin zone** is a primitive unit cell of the reciprocal lattice.

Thus real lattice indicates the location of real objects (atoms) and has dimension of *m* whereas the reciprocal lattice indicates the positions of abstract points (magnitude and direction of momentum) and has dimension of m^{-1}.

CONSTRUCTION OF TWO-DIMENSIONAL RECIPROCAL LATTICE

Choose any point in the direct lattice as an origin. Then

1. from this origin, lay out the normal to every family of parallel planes in the direct lattice.

2. set the length of each normal equal to 2π times the reciprocal of the interplanar spacing for its particular set of planes.

3. place a point at the end of each normal.

For example, a two-dimensional real lattice defined by two unit cell vectors a and b inclined at an angle γ is shown in the Figure 4.1. The equivalent reciprocal lattice in reciprocal space is defined by two reciprocal vectors, a^* and b^*.

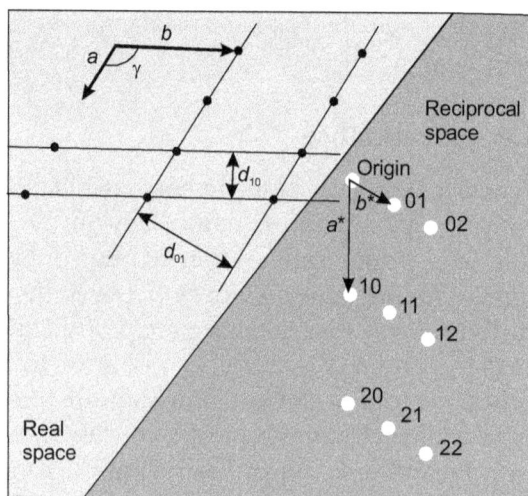

Figure 4.1   Two-dimensional real and reciprocal lattice

The reciprocal vectors are defined as follows:

- a^* is of magnitude $1/d_{10}$, where d_{10} is the spacing of the (10) planes, and is perpendicular to b

- b^* is of magnitude $1/d_{01}$, where d_{01} is the spacing of the (01) planes, and is perpendicular to a.

The reciprocal lattice can be described by just one unit cell, which can be multiplied by translation along all the coordinate axes in the same fashion as a direct lattice. The knowledge of the reciprocal lattice is accomplished using the so called **Ewald construction** which allows one to put the information about the wavelength and the direction of the incident radiation into the reciprocal lattice and determines the diffraction pattern in a relatively straightforward way. At the surface, the translational symmetry of the bulk of a crystal breaks down. A surface has finite translations in two dimensions, which lay in its plane, and the absence of a translation along the normal can be envisaged as a translation to infinity in that direction. The reciprocal lattice and the Ewald construction of the surface are built following the same rules as for the bulk. However, since the reciprocal of infinity is equal to zero, it has a multitude of points laying infinitely close to each other in the direction of the normal. The collection of these points, which spread from the two-dimensional surface lattice to infinity, is called Bragg's rods. Since the rods are infinitely dense lattice points, the diffraction from a surface occurs continuously with changes in the direction and the magnitude of the incident wave vector as long as the wave is short enough to be diffracted.

THE RECIPROCAL OF A BRAVAIS LATTICE

A set of wave vectors K (Figure 4.2) that specify the periodicity of a Bravais lattice of sites R will be useful in discussing X-ray scattering of ions and electron eigen states in ionic potential.

Suppose we have a function $U(r)$ that is periodic on the Bravais lattice, i.e., we have

$$U(r + R) = U(r)$$

for all R in the Bravais lattice. Taking the Fourier's transform

$$U(r) = \int \frac{d^3K}{(2\pi)^3} e^{iK.r} U(K)$$

the above condition becomes

$$U(r + R) = \int \frac{d^3K}{(2\pi)^3} e^{iK.(r+R)} U(K) \qquad (4.4)$$

where, $e^{iK.(r+R)} = e^{iK.r}$ for all R in Bravais lattice.

Therefore,

$$U(r + R) = \int \frac{d^3K}{(2\pi)^3} e^{iK.r} U(K)$$

$$= U(r)$$

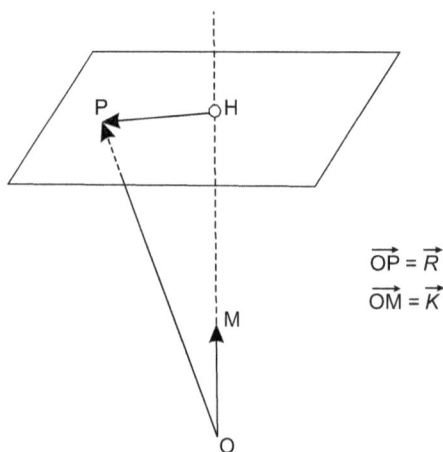

$$\overrightarrow{OP} = \overrightarrow{R}$$
$$\overrightarrow{OM} = \overrightarrow{K}$$

Figure 4.2   Real and reciprocal lattice vector

If this is to be true, then only the values of K for which $U(K) = 0$ must be the set of K such that $e^{iK.R} = 1$ for all

R in the Bravais lattice. This defines the reciprocal lattice K. Alternatively, the set of wave vectors K that yield plane waves with the periodicity of the Bravais lattice is called the reciprocal lattice.

As plane wave is invariant under translation by R, $e^{iK \cdot r} = 1$ for all R in Bravais lattice.

The Reciprocal Lattice is itself a Bravais Lattice

The basic set of the reciprocal lattice vectors are defined by the equation

$$a_i \cdot \vec{a_j^*} = 2\pi\delta_{ij} \{\delta_{ij} = 0 \text{ if } i \neq j; \delta_{ij} = 1 \text{ if } i = j\}$$

The a_i's are the fundamental vectors of the direct lattice which means we can write any wave vector R as a linear combination.

$$R = b\vec{a_1} + k\vec{a_2} + l\vec{a_3}$$

The a_i^*'s are then called the fundamental vectors of the reciprocal lattice. Then for any K in the reciprocal lattice

$$K = x\vec{a_1^*} + y\vec{a_2^*} + z\vec{a_3^*}$$

$$\begin{aligned} K \cdot R &= (b\vec{a_1} + k\vec{a_2} + l\vec{a_3}) \cdot (x\vec{a_1^*} + y\vec{a_2^*} + z\vec{a_3^*}) \\ &= 2\pi(bx + ky + lz) \end{aligned} \quad (4.5)$$

If K is in the reciprocal lattice, we must have $e^{iK \cdot R} = 1$ for all R.

RECIPROCAL LATTICE OF VARIOUS CRYSTAL SYSTEMS

The reciprocal lattices for the cubic crystal system are as follows:

Simple Cubic Lattice

Let us consider a simple cubic Bravais lattice, with cubic primitive cell of side a ($a_1 = a_2 = a_3 = a$) and whose direct lattice vectors are

$$\vec{a}_1 = a\hat{x}$$
$$\vec{a}_2 = a\hat{y} \qquad (4.6)$$
$$\vec{a}_3 = a\hat{z}$$

and for its reciprocal, a simple cubic lattice with a cubic primitive cell of side $2\pi(1/a)$, in the crystallographer's definition. Then its reciprocal lattice vectors are

$$\vec{a_1^*} = \frac{2\pi}{a}\hat{x}$$
$$\vec{a_2^*} = \frac{2\pi}{a}\hat{y} \qquad (4.7)$$
$$\vec{a_3^*} = \frac{2\pi}{a}\hat{z}$$

Thus reciprocal lattice of a simple cubic crystal system is also a simple cubic of length $2\pi/a$.

Face Centred Cubic Lattice

Consider the direct lattice vectors of a face centred cubic lattice as

$$\vec{a}_1 = \frac{a}{2}(\hat{y} + \hat{z})$$
$$\vec{a}_2 = \frac{a}{2}(\hat{z} + \hat{x}) \qquad (4.8)$$
$$\vec{a}_3 = \frac{a}{2}(\hat{x} + \hat{y})$$

And by construction $\overrightarrow{a_i^*}$ to get

$$\overrightarrow{a_1^*} = \frac{2\pi}{a}(\hat{y} - \hat{x} + \hat{z})$$

$$\overrightarrow{a_2^*} = \frac{2\pi}{a}(\hat{z} - \hat{y} + \hat{x})$$

$$\overrightarrow{a_3^*} = \frac{2\pi}{a}(\hat{x} - \hat{z} + \hat{y})$$

(4.9)

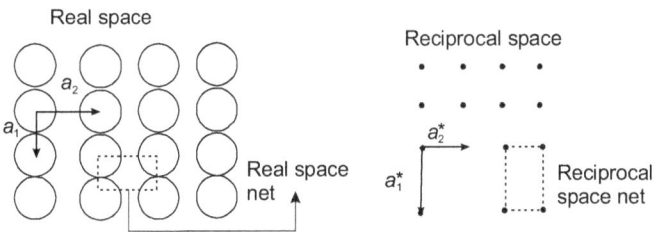

Figure 4.3 Two-dimensional reciprocal lattice for fcc (110)

These $\overrightarrow{a_1^*}, \overrightarrow{a_2^*}$ and $\overrightarrow{a_3^*}$, are the primitive vectors of a bcc lattice with side of the cubic unit cell equal to $4\pi/a$. The two-dimensional reciprocal lattice plane (110) for fcc is shown in Figure 4.3.

Body Centred Cubic Lattice

Since the reciprocal of the reciprocal lattice is the direct lattice, we can say the reciprocal of the bcc lattice is an fcc lattice. If the bcc direct lattice has a unit cubic cell of length a, then the reciprocal (fcc) lattice has a unit cubic cell of length $4\pi/a$.

Hexagonal Bravais Lattice

The reciprocal of the single hexagonal Bravais lattice with lattice constants a and c is also a single hexagonal lattice with lattice constants $\left|\overrightarrow{a_1^*}\right| = \left|\overrightarrow{a_2^*}\right| = 4\pi/\sqrt{3}a$ and $\left|\overrightarrow{a_3^*}\right| = \frac{2\pi}{c}$.

THE EWALD SPHERE

A geometrical description of diffraction was originally proposed by P.P. Ewald. The advantage of this description is that it allows the determination of which Bragg reflections will be observed, knowing the orientation of the crystal with respect to the incident beam. The Ewald construction provides a geometrical relationship between the orientation of the crystal and the direction of the X-ray beams diffracted by it. The Bragg's condition for diffraction occurs when a set of lattice planes, with defined d_{hkl} spacing, are inclined with respect to the incident beam by an angle θ. The diffracted beam (Bragg reflection) occurs at 2θ from the incident beam. The diffraction vector (defined as $H = (S-S_0)/\lambda$) is perpendicular to the lattice planes.

The crystal (C) can be physically oriented so that a required reciprocal lattice point can intersect the sphere. From there, S_0 (the direction of the diffracted beam) can be deduced from $H = (S-S_0)/\lambda$ (Figure 4.4a).

In the Ewald construction, a sphere with radius $1/\lambda$ is drawn. The reciprocal lattice, drawn at the same scale as that of the Ewald sphere, is then placed with its origin centred at O (Figure 4.4b). As the crystal is rotated, the crystal lattice and also the reciprocal lattice rotates. If during the rotation of the crystal a reciprocal lattice point (*hkl*) touches the surface of the sphere, Bragg's law is satisfied. The result is a reflection in the direction S_0, with values of *h*, *k*, *l* corresponding both to the values of the reciprocal lattice point and for the crystal lattice planes.

The X-ray diffraction pattern of a crystal is the sampling at the reciprocal lattice points of the X-ray diffraction pattern of the contents of a single unit cell. It is only necessary to find the atomic arrangement in one unit cell, which can be

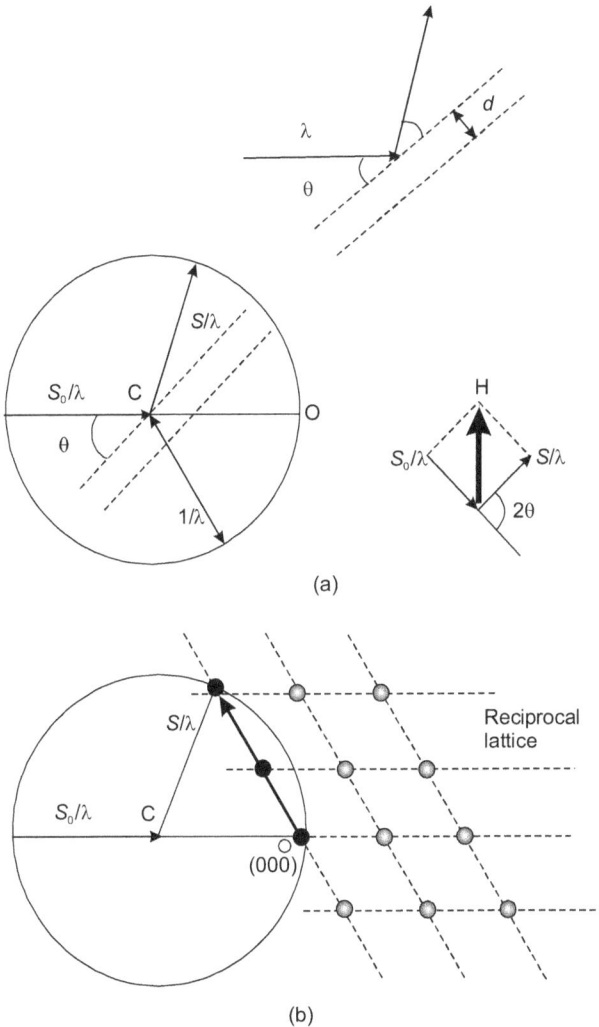

Figure 4.4 Construction of the Ewald sphere, illustrated in two dimensions (Ewald circle). a) Bragg's law and definition of S and S_0, construction of the sphere (radius $= 1/\lambda$) centred on the crystal. b) Orientation of the reciprocal lattice with its origin ($hkl = 000$) at O.

derived from the overall intensity variation in the diffraction pattern. This atomic arrangement is then repeated according to the direct lattice to give the entire crystal structure. The spatial arrangement of the diffracted beams is determined by the geometry of the crystal lattice while the intensities are determined by the arrangement of atoms within one unit cell.

X-RAYS AND THE PRODUCTION OF X-RAYS

X-rays are electromagnetic radiation with wavelengths between 0.02 Å and 100 Å (1 Å = 10^{-10} metres). They are part of the electromagnetic spectrum (Figure 4.5) that includes wavelengths of electromagnetic radiation called visible light which our eyes are sensitive to (different wavelengths of visible light appear to us as different colours). *Because X-rays have wavelengths similar to molecular dimensions, they are useful to explore these within crystals.*

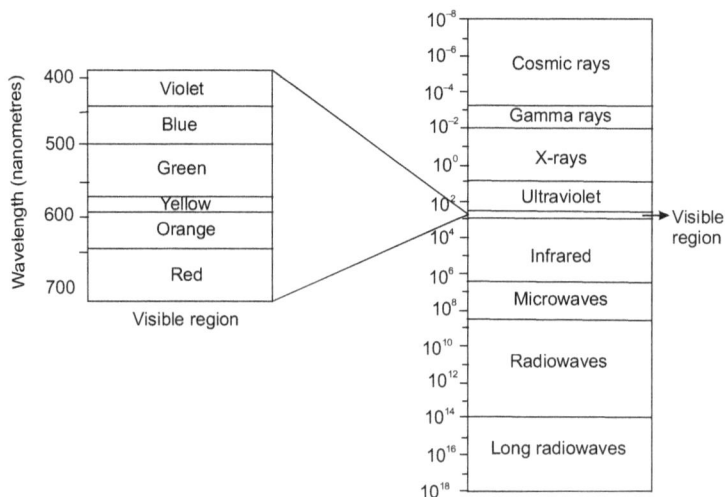

Figure 4.5   Electromagnetic spectrum

The energy of X-rays, like all electromagnetic radiation, is inversely proportional to their wavelength as given by the Einstein equation

$$E = h\nu = hc/\lambda$$

where,

  E = energy,
  h = Planck's constant = 6.62517×10^{-27} erg sec.,
  ν = frequency,
  c = velocity of light = 2.99793×10^{10} cm/s and
  λ = wavelength.

Thus, since X-rays have a smaller wavelength than visible light, they have higher energy. With their higher energy, X-rays can penetrate into matter more easily than visible light. Their ability to penetrate matter depends on the density of the matter, and thus X-rays provide a powerful tool in medicine for mapping internal structures of the human body (bones have higher density than tissue, and thus are harder for X-rays to penetrate, fractures in bones have a different density than the bone, thus fractures can be seen in X-ray pictures).

X-rays are produced in a device called X-ray tube. Such a tube is illustrated in Figure 4.6. It consists of an evacuated chamber with a tungsten filament at one end of the tube, called the cathode, and a metal target at the other end, called an anode. Electrical current is run through the tungsten filament, causing it to glow and emit electrons. A large voltage difference (measured in kilovolts) is placed between the cathode and the anode, causing the electrons to move at high velocity from the filament to the anode target. Upon striking the atoms in the target, the electrons dislodge inner shell electrons resulting in outer shell electrons to jump to a lower energy shell replacing the dislodged electrons. These

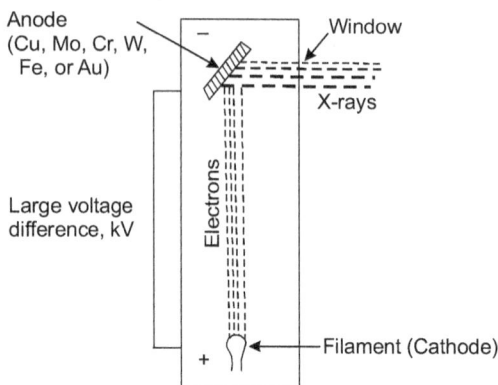

Figure 4.6   X-ray tube

electronic transitions result in the generation of X-rays. The X-rays then move through a window in the X-ray tube and can be used to provide information on the internal arrangement of atoms in crystals or the structure of internal body parts.

When the target material of the X-ray tube is bombarded with electrons accelerated from the cathode filament, two types of X-ray spectra are produced. The first is called the **continuous spectra**.

The continuous spectra consist of a range of wavelengths of X-rays (Figure 4.7) with minimum wavelength and intensity (measured in counts per second) dependent on the target material and the voltage across the X-ray tube. When the applied voltage increases, the minimum wavelength decreases and intensity tends to increase. The second type of spectra, called the **characteristic spectra**, is produced at high voltage as a result of specific electronic transitions that take place within individual atoms of the target material.

This is the easiest to see using the simple Bohr model of the atom. In this model, the nucleus of an atom containing

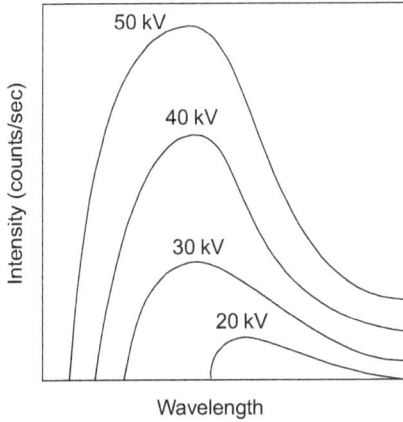

Figure 4.7 Continuous X-ray spectra

the protons and neutrons is surrounded by shells of electrons. The innermost shell, called the K-shell, is surrounded by the L- and M-shells. When the energy of the electrons accelerated towards the target becomes high enough to dislodge K-shell electrons, electrons from the L- and M-shells move to the K-shell to replace the dislodged electrons. (Figure 4.8).

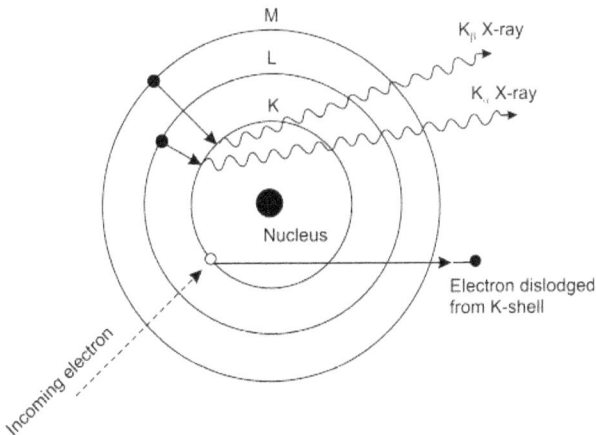

Figure 4.8 Generation of X-rays

Each of these electronic transitions produces X-ray with a wavelength that depends on the exact structure of the atom being bombarded. A transition from the L-shell to the K-shell produces a K_α X-ray, while the transition from an M-shell to the K-shell produces a K_β X-ray.

These characteristic X-rays have a much higher intensity than those produced by the continuous spectra, with K_α X-rays having higher intensity than K_β X-rays (Figure 4.9). The important point here is that the wavelength of these characteristic X-rays is different for each atom in the periodic table (of course only those elements with higher atomic number have L- and M-shell electrons that can undergo transitions to produce X-rays). A filter is generally used to filter out the lower intensity K_β X-rays.

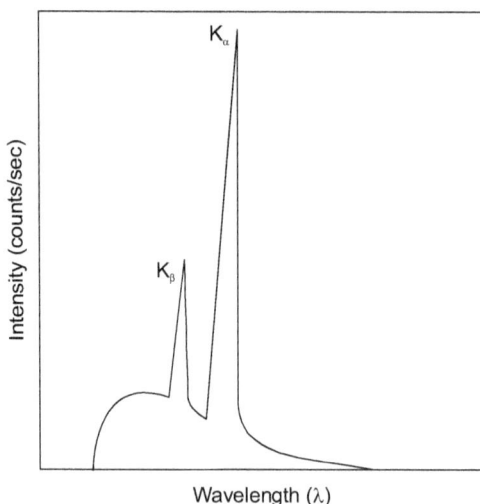

Figure 4.9 K_α and K_β radiations

For commonly used target materials in X-ray tubes, the X-rays have the following well known experimentally determined wavelengths.

Element K$_\alpha$	Wavelength (λ) Å
Mo	0.7107
Cu	1.5418
Co	1.7902
Fe	1.9373
Cr	2.2909

WHY X-RAY DIFFRACTION?

The basic steps that occur in all processes of image formation are as follows: (a) scattering of the radiation and (b) recombination of the scattered beams. The basic idea of an ordinary 2" × 2" slide projector is if the lens is removed so that a diffuse patch of light is seen on the screen even though a slide is in the same position, it will be clear to audience that all the information that is contained in the slide must be available in the patch of light on the screen although it is not readily decipherable. Clearly the lens cannot "know" anything about the slide and yet as soon as it is placed in the correct position, the nature and details of the slide are revealed. All that the lens can do is to rearrange the information so that it is immediately understandable to the eye and brain.

With visible light we can easily solve the focusing problem and images of extremely small objects may be produced in the optical microscope. One severe limitation however, is the wavelength of light and details below this size cannot be imaged. One alternative is to use electrons whose wavelength is quite small enough, but the practical problems of lens designs for the electron microscope provide an experimental limit before the resolution of individual atoms can be achieved.

X-rays have a suitable wavelength and would provide a simple solution if they could be focused experimentally. Unfortunately this is not possible except with systems of curved mirrors which are capable of only very limited magnification. To achieve the full benefits of the small wavelength, some alternative approach must be adopted.

In principle one could say that the whole development of X-ray diffraction techniques really amounts to the development of alternatives to the focusing of X-ray images. The point is that the first stage of the imaging process, illustrated by the projector with no lens, can be performed but the crystallographer has no lens to put back in the projector and must try to make sense out of the diffuse patch in some other way.

If the problem were strictly analogous to this, it is unlikely that any structure would ever have been solved. Fortunately there are two significant ways in which the X-ray crystallographer's case differs from that of the projectionist with no lens for his projector. First of all, the projector uses white light with a broad frequency band, which is also spatially incoherent and is produced from a large source. In the case of X-ray it is usual (except under the special circumstances of Laue photographs with which we are not concerned here) to use monochromatic radiation, which, as a result of travelling through a long, fine hole or slit has quite a high degree of spatial coherence. The second point is that the object usually exhibits some degree of regularity or crystallinity.

These two facts lead to the production of patterns, which consist of not a diffuse patch but rather a series of discrete spots.

The process carried out by the lens of the projector or by the objective of the microscope and which needs to be carried out artificially by the X-ray crystallographer involves the mathematical operation of **Fourier synthesis**. If the scattering (or diffraction pattern as it tends to be called if it consists of regular spots) is completely determined, then it should, in principle, be possible to transform the beam into an image by the purely mathematical process of Fourier synthesis. Unfortunately, however, it has so far been proved quite impossible to record the relative phases; this immediately invalidates the direct mathematical process. The reason why the phase cannot be recorded becomes clear if one calculates the frequency of X-rays; determination of phase would, in effect, involve time measurements that corresponding to a fraction of one period. *If we assume that the wavelength of X-ray is 1.5 Å (1.5 × 10⁻¹⁰ m), the frequency is about 2 × 10¹⁸ hertz and hence to measure a phase difference of (say) 1/5th of a cycle would involve a time measurement of about 10⁻¹⁹ s which is certainly beyond our present resources.* Perhaps one day a means of adding a coherent beam, as in optical-laser holography may become available and then the whole situation would change!

ELECTRON IN A PERIODIC POTENTIAL

In an ideal crystal, the ions occupy positions, which form a regular periodic structure. The potential $U(r)$ is thus a periodic function with the period equal to that of the corresponding Bravais lattice:

$$U(r + R) = U(r) \qquad (4.10)$$

where, r and R are the vectors which belong to the Bravais lattice. The period of the potential is of the same order as the de Broglie wavelength which requires quantum-mechanical consideration of the problem. As the total Hamiltonian for solids contain electron–electron interaction terms, the problem represents the many-body system. Within the theory of independent electrons, an effective single-electron potential $U(r)$ is introduced. In the case of the ideal periodic crystal this potential must satisfy the equation 4.10.

The main purpose is to analyse the periodicity-induced properties of the single-electron Schrödinger equation

$$\left(-\frac{\hbar^2}{2m}\nabla^2 + U(r)\right)\psi(r) = E\psi(r) \tag{4.11}$$

Due to the potential periodicity, the solution to this equation has several remarkable properties briefly given below.

Statement of the Theorem   Let R be any vector in a lattice. Let ψ be a single electron solution to the Schrödinger equation

$$\left(-\frac{\hbar^2}{2m}\right)\partial^2\psi/\partial x^2 + U(r)\psi(r) = E\psi(r) \tag{4.12}$$

where, $U(r + R) = U(r)$ for all lattice vectors R. Therefore there exists a wave vector k in the reciprocal lattice and a periodic function $U_k(r)$ such that $U_k(r + R) = U_k(r)$. Then ψ is of the form

$$\psi(r) = e^{\pm ik.r}U_k(r) \tag{4.13}$$

This means that ψ is a plane wave $e^{ik.r}$ modulated by the function $U_k(r)$. This theorem is known as Bloch theorem.

Not all wave functions satisfy the Bloch theorem. For example, if the wavefunction is for a lattice with boundaries then it is not of the Bloch form. The wavefunction of two or more interacting electrons is not of the Bloch form.

If $\psi_k(x) = C(k)e^{ik.x} + C(k + G)e^{i(k + G).x}$

then

$$\psi_k(x) = e^{ik.x}[C(k) + C(k+G)e^{iG.x}] \tag{4.14}$$

Consider if,

$$U_k(x) = [C(k) + C(k + G)e^{iGx}]$$

is periodic with the periodicity of the lattice. Let R be any element of the lattice and let G be any element of the reciprocal lattice. This means that GR is an integral multiple of 2π so

$$e^{iGR} = 1$$

Then

$$U_k(x + R) = [C(k) + C(k + G)\exp(iG(x + R))]$$

$$= [C(k) + C(k+G)\exp(iGx)\exp(iGR)] \quad (4.15)$$

But $e^{iGR} = 1$ so

$$U_k(x+R) = [C(k) + C(k+G)e^{iGx}] = U_k(x) \quad (4.16)$$

Thus $\psi_k(x)$ is of the Bloch form.

$\psi_k(x)$ will be an eigen function of the momentum operator if the momentum operator commutes with the Hamiltonian operator. This requires that

$$\partial/\partial x(H) = H(\partial/\partial x); \text{ If } f(x) = 0 \quad (4.17)$$

for any function $f(x)$.

While it is true that

$$(\partial/\partial x)(\partial^2/\partial x^2) = (\partial^2/\partial x^2)(\partial/\partial x) = (\partial^3/\partial x^3) \quad (4.18)$$

DIFFRACTION OF X-RAYS BY CRYSTAL LATTICE

Coherent Scattering of X-rays by Electrons

To discuss diffraction of X-rays by atoms in a crystal, it is convenient to consider the wave nature of X-ray radiation. Recall that all electromagnetic radiation can be described by sinusoidally varying electric and magnetic fields. The electric field of X-ray beam incident on an electron interacts with the electric charge of the electron and is scattered in all directions. Although the nucleus of an atom also has an electric charge, it is relatively massive and does not interact strongly with X-rays. Thus, X-rays are scattered from atoms almost exclusively by interactions with their electrons. There are two scattering mechanisms. Coherently scattered X-rays have the same wavelength as the incident radiation, and have electric and magnetic fields that are "in-phase" with the incident radiation. The intensity of coherently scattered X-rays varies with direction in the manner described by the

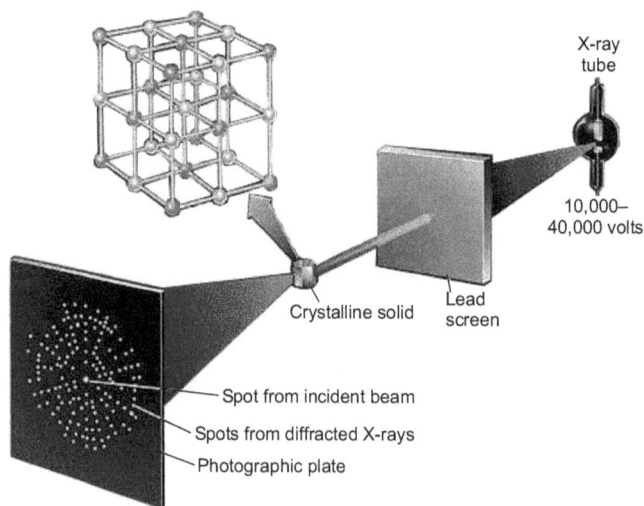

Figure 4.10   Diffraction of X-rays by a crystal

Thomson equation, which reveals that the scattered X-rays have the highest intensity in the forward and reverse directions, and have the lowest intensity at right angles to the incident X-ray beam. As a result of their well defined phase relationship with the incident X-ray beam, some of the coherently scattered X-rays eventually form the diffracted beams that we detect in X-ray diffraction experiments (Figure 4.10).

Scattering by One Atom

The preceding paragraph describes coherent scattering of X-rays by individual electrons. Since the number of electrons in an atom depends on the atomic number (Z), we might expect that the intensity of X-rays scattered from an atom will be proportional to Z. This is true in the forward and reverse scattering directions, but is not quite correct for other scattering directions. The additional factor that must be considered for scattering at an angle to the incident X-ray beam arises because the electrons surrounding the nucleus of an atom are located at different positions. This further decreases the scattered intensity for scattering directions that make an angle $[\theta]$ with the incident X-ray beam, beyond the direction-dependent scattering intensity from individual electrons as described by the Thomson equation

$(I_e = I_o \dfrac{e^4}{r^2 m^2 c^4} \left(\dfrac{1 + \cos^2 \phi}{2} \right)$, where I_c is the intensity of coherent scattering by one electron). However, for a given scattering angle $[\theta']$ relative to the incident X-ray beam, a single atom of Fe ($Z = 26$) should scatter X-rays with about twice the intensity of a single atom of Al ($Z = 13$).

Diffraction from a One-Dimensional Crystal

If every atom scatters X-rays in all directions, why don't we observe X-rays being scattered from a crystal in all directions? The answer lies in the periodic arrangement of atoms in crystals. The scattered X-rays suffer destructive interference in most directions, giving zero scattered intensity. They experience constructive interference in a few directions, giving strong diffracted beams. A diffracted X-ray beam is a beam of X-rays scattered from all the atoms in a crystal such that the scattered beams are in phase and therefore mutually reinforce one another. Since this effect arises from the periodic arrangement of atoms in a crystal, we next examine the derivation of Bragg's law—the rule that identifies the directions of diffracted X-ray beams.

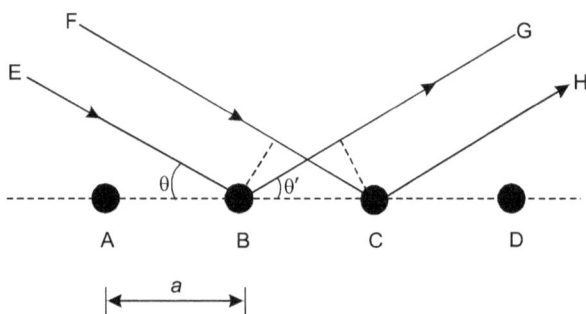

Figure 4.11 Diffraction from a single row of atoms (1-dimensional crystal)

To understand the concept of diffraction, we can study scattering by a pair of atoms that could be part of a longer row of atoms (a 1-dimensional crystal). The atoms of the pair are labelled B and C in Figure 4.11. The path length difference for X-rays scattered from these 2 atoms is zero (FC + CH = EB + BG), if the scattering angle θ' equals the incidence angle θ. Because the X-rays scattered from atoms B and C travel the same distance, they will be in phase and will constructively interfere. The result is a diffracted beam at $\theta' = \theta$. However, further consideration shows that the scattered X-rays can be in phase at other scattering angles too. What is required for X-rays scattered from atoms A and D to be in phase is that the path lengths travelled by the X-rays scattered from atoms A and D must differ by an integral number of wavelengths. Once we see this, it is straightforward to construct a geometrical analysis that yields an equation for the directions of the X-ray beams that are in phase after being scattered from a pair of atoms. The result is

$$a\cos\theta + n\lambda = a\cos\theta' \qquad (4.19)$$

where, a is the atom spacing (lattice parameter in a 1-D crystal) and n is a positive or negative integer. If the 1-D crystal is now considered to consist of a long row of atoms instead of just 2 atoms, the requirement that the scattered X-rays be in phase is the same as before. This is because each pair of adjacent atoms in the 1-dimensional crystal produces diffracted beams at the angles specified by equation 4.19. If the scattered X-rays are in phase for the nearest neighbour atoms, they must also be in phase for the second nearest neighbour atoms, etc. Thus, equation 4.19 is the rule for determining the direction of diffracted X-ray beams for a 1-dimensional crystal.

Before proceeding, it is worth considering how many lattice spacings apart the atoms can be and still produce the constructive interference that is required to form a diffracted beam. The answer depends on many factors such as the divergence of the incident X-ray beam and the size of the detector window. For practical X-ray diffraction experiments, X-rays scattered from all atoms in a crystal contribute to the formation of diffracted beams.

Although 1-dimensional crystals (isolated polymer chains for example) are not easy to work with, we can use a macroscopic analog to demonstrate the diffraction principle for a 1-dimensional crystal. A diffraction grating is a series of fine lines scribed on a transparent substrate, with spacing slightly larger than the wavelength of visible light. Since the lines go only in one direction, they constitute a 1-dimensional array of scattering centers or a 1-dimensional crystal in materials science language. The spacing of the lines a is equivalent to the lattice parameter of the 1-dimensional crystal. Instead of X-rays, we can use laser as a source of monochromatic (one wavelength, just like the characteristic X-ray line) radiation. If we shine the laser beam on the diffraction grating with a normal incidence ($\theta = 90°$), then equation 4.19 is simplified to

$$n\frac{\lambda}{a} = \cos\theta' \qquad (4.20)$$

where, n is a positive or negative integer and

a is the spacing of the lines on the grating.

This equation shows that we should see a series of diffracted light beams at various angles θ' that depend on the ratio λ/a (which is a constant). This can be extended to the two-dimensional case as well.

LAUE'S FORMULATION OF X-RAY DIFFRACTION

Since a crystal is a regular three-dimensional arrangement of unit cells, it can be regarded as a three-dimensional diffraction grating for X-rays. The effect of a grating is to limit the directions in which an observable diffracted beam occurs; in the diffraction of X-rays by crystals it will be shown that the directions of the directed X-ray beams depend on the dimensions of the unit cell and their intensities on the nature and deposition of atoms within the unit cell.

By analogy with the corresponding points in the grating element we can take corresponding points, one in each unit cell of the crystal and so obtain an array of lattice points. Diffracted beams will occur in directions for which X-rays scattered by all lattice points are in phase and, since the lattice is a regular three-dimensional array of lattice points, this condition is satisfied by pairs of adjacent lattice points lying on three non-coplanar rows.

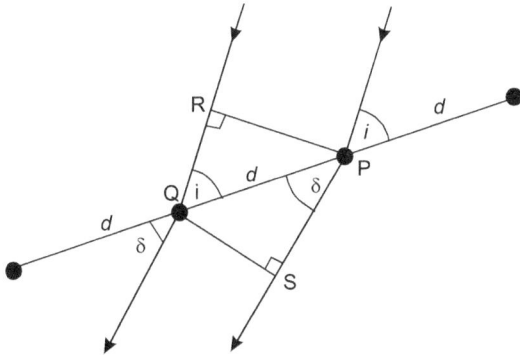

Figure 4.12 A parallel beam of X-rays of wavelength λ, represented by the wavefront PR, being incident at an angle of incidence i on a row of lattice points, P, Q,... of separation d. The condition for the X-rays scattered by adjacent lattice points to be in phase is PS–RQ $= n\lambda$, where n is an integer.

Let us consider a parallel beam of X-rays of wavelength λ to be incident on a row of lattice points of spacing d at an angle of incidence i (Figure 4.12) and consider a direction of scattering at an angle δ to the lattice row. The path difference for X-rays scattered by adjacent lattice points with reference to Figure 4.12 is

$$PS - RQ = PQ(\cos \delta - \cos i)$$
$$= d\,(\cos \delta - \cos i) \qquad (4.21)$$

For a diffracted beam to occur, the X-rays scattered by adjacent lattice points have to be in phase, that is to say their path difference must be an integral number of wavelengths. Therefore the condition for diffraction is

$$d(\cos \delta - \cos i) = n\lambda$$

where, n is an integer.

The permissible directions of the diffracted beam are of course not confined to the plane defined by the incident

beam and the *lattice point row*. The condition for diffraction
$d (\cos \delta - \cos i) = n\lambda$, is satisfied by any direction making
the angle δ with the lattice row. Therefore for a given value
of n the diffracted radiation is confined to the surface of a
cone of semi angle δ; a set of cones coaxial about the lattice
row represents solutions of the diffraction condition for
$n = 0, +1, +2$, etc. (Figure 4.13).

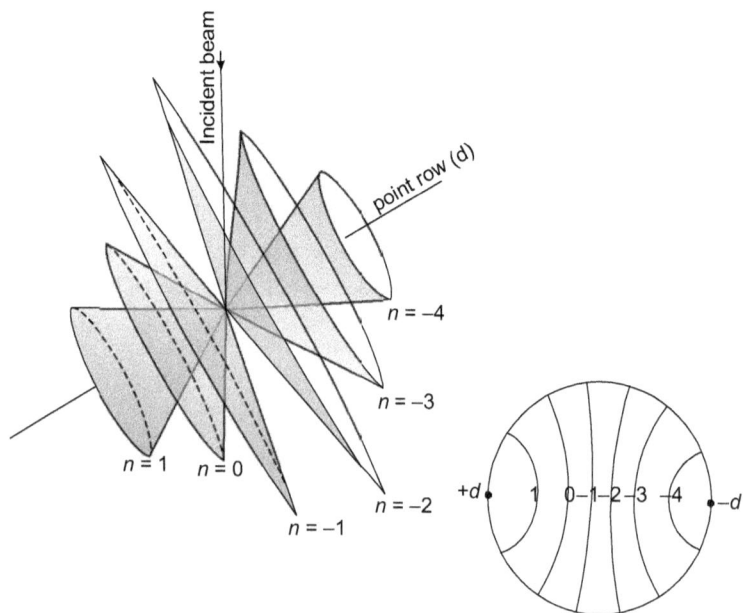

Figure 4.13 A set of cones, each corresponding to a particular value
of n in the equation $d(\cos \delta - \cos i) = n\lambda$, coaxial
about a selected lattice row represents solutions of the
diffraction equation for the lattice row. The cones of
diffracted intensity maxima are shown in perspective and
in stereographic projection.

So far we have considered diffraction by a single point
row, but a lattice is a regular three-dimensional array of
points and as such is completely specified by the distance

apart of adjacent lattice points in three non-coplanar directions. Let us consider these axes as x, y, and z and the separation of lattice points along each to be a, b, and c respectively. For a row of separation, a, parallel to the x-axis, we have seen that the diffraction condition is

$$a(\cos \delta_a - \cos i_a) = h\lambda$$

where, i_a and δ_a are the angles between the x-axis and the incident and diffracted beams, respectively, and h is an integer.

If such a set of row is repeated successively with translation, b, parallel to the y-axis, each row becomes a grating element and the diffraction condition for this grating is

$$b(\cos \delta_b - \cos i_b) = k\lambda$$

where, k is an integer.

The condition for the production of an observable diffracted beam by all the lattice points in the xy-plane is then $a(\cos \delta_a - \cos i_a) = h\lambda$ and $b(\cos \delta_b - \cos i_b) = k\lambda$ simultaneously. If such a set of lattice planes is repeated successively with translation c parallel to the z-axis, each plane constitutes an element of a third linear diffraction grating for which the diffraction condition is $c(\cos \delta_c - \cos i_c) = l\lambda$, where l is an integer. The simultaneous operation of all three conditions is necessary for the production of an observable diffracted beam by all the lattice points of the three-dimensional lattice. This statement implies that if X-rays scattered by any pair of adjacent lattice points in a three-dimensional lattice are to be in phase, then the X-rays scattered by adjacent lattice points along each of the reference axes must be in phase; this constitutes a necessary and sufficient condition for diffraction by a three-dimensional array of lattice points.

The diffraction condition for a three-dimensional lattice can thus be written as

$$a(\cos \delta_a - \cos i_a) = h\lambda$$
$$b(\cos \delta_b - \cos i_b) = k\lambda$$
$$c(\cos \delta_c - \cos i_c) = l\lambda \qquad (4.22)$$

where,

the incident beam is inclined at angles i_a, i_b, i_c and the diffracted beam at angles δ_a, δ_b, δ_c with the x, y, z axes respectively,

λ is the X-ray wavelength, and

h, k, l are integers.

These three equations are known as the **Laue equations** named after Max von Laue, who in 1912, suggested that a crystal should act as a diffraction grating of X-rays.

The first Laue equation, $a (\cos \delta_a - \cos i_a) = h\lambda$, restricts the directions of observable diffracted beams to the surfaces of a set of cones coaxial about the x-axis and having the semiangles δ_a consistent with the equation. For a given direction i of the incident beam, the restriction placed by the first Laue equation on the directions of diffracted beams can be represented by a set of small circles of radius δ_a centred on the x-axis.

The second Laue equation, $b(\cos \delta_b - \cos i_b) = k\lambda$, further constraints the directions of observable diffracted beams to a set of small circles of radius δ_b centred on the y-axis. The simultaneous operation of these two Laue equations thus restricts observable diffracted beams produced by X-rays incident in a particular direction to the intersection of pairs of small circles, the attitude of one small circle being dependent on h and of the other on k. Each pair of h, k values leads, in general, to two common directions.

The third Laue equation, $c(\cos \delta_c - \cos i_c) = l\lambda$, provides an additional restriction; for an observable diffracted beam to occur, a small circle of the sets of radius δ_c centred on the z-axis must pass through the intersection of small circles of the x and y sets. An observable diffracted beam thus lies in a direction common to three cones each coaxial with one of the reference axes and of semiangle consistent with the appropriate Laue equation; such a direction is completely specified for a given lattice and for X-rays of given λ by the integers h, k and l. The first of these solutions of the Laue equations are trivial, representing merely the forward direction of the incident beam.

It is worth noticing at this point that the three-dimensional grating, such as a crystal, differs from a one- or two-dimensional grating in the small number of diffracted beams produced by any particular orientation of the incident beam. Other diffracted intensity maxima can be observed only by changing the orientation of the incident beam relative to the crystallographic reference axes; in practical X-ray crystallography, this is most conveniently done by rotating the crystal and keeping the orientation of the incident beam fixed.

X-RAY DIFFRACTION AND BRAGG'S LAW

Since a beam of X-rays consists of a bundle of separate waves, the waves can interact with one another. Such interaction is termed interference. If all the waves in the bundle are in phase, that is their crests and troughs occur at exactly the same position, the waves will interfere with one another and their amplitudes will add together to produce a resultant wave (Figure 4.14) that has a higher amplitude (the sum of all the waves that are in phase).

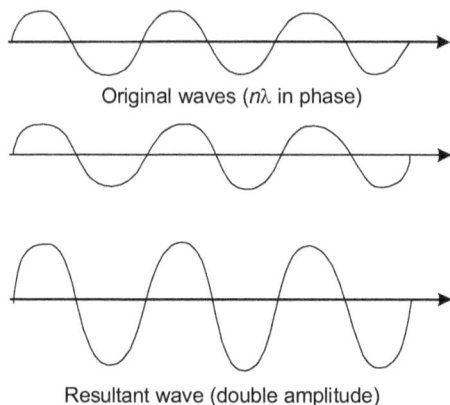

Figure 4.14   Interference between two waves ($n\lambda$ in phase)

If the waves are out of phase, being off by a non-integral number of wavelengths, then destructive interference will occur and the amplitude of the waves will be reduced. In an extreme case, if the waves are out of phase by a multiple of $\frac{1}{2}\lambda$, the resultant wave will have no amplitude and thus be completely destroyed (Figure 4.15).

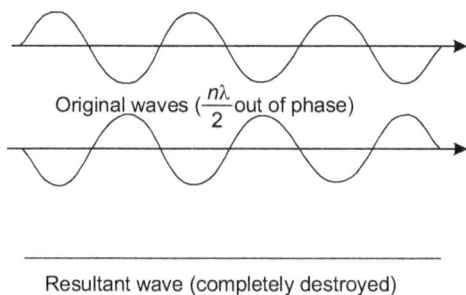

Figure 4.15   Interference between two waves ($\frac{n\lambda}{2}$ out of phase)

The atoms in crystals interact with X-ray waves in such a way as to produce interference. The interaction can be thought of as if the atoms in a crystal structure reflect the waves. But, because a crystal structure consists of an orderly arrangement of atoms, the reflections occur from what appears to be planes of atoms. Let us imagine a beam of X-rays entering a crystal with one of these planes of atoms oriented at an angle of θ to the incoming beam of monochromatic X-rays (monochromatic means one colour, or in this case 1 discrete wavelength as produced by the characteristic spectra of the X-ray tube).

Two such rays are shown here, where the spacing between the atomic planes occurs over the distance, d. Ray 1 reflects off from the upper atomic plane at an angle θ equal to its angle of incidence (Figure 4.16). Similarly, Ray 2 reflects off from the lower atomic plane at the same angle θ. While ray 2 is in the crystal, however, it travels a distance of $2a$ farther than ray 1. If this distance $2a$ is equal to an integral number of wavelengths ($n\lambda$), then rays 1 and 2 will be in phase on their exit from the crystal and constructive interference will occur.

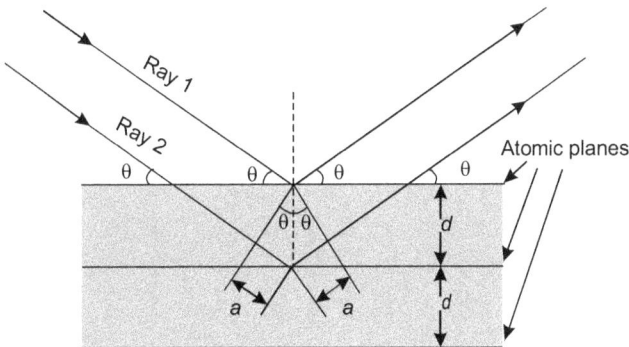

Figure 4.16 Bragg's Law

If the distance $2a$ is not an integral number of wavelengths, then destructive interference will occur and the waves will not be as strong as when they entered the crystal. Thus, the condition for constructive interference to occur is

$$n\lambda = 2a$$

but, from trigonometry, we can figure out what the distance $2a$ is in terms of the spacing, d, between the atomic planes.

$$a = d \sin \theta$$
$$\text{or } 2a = 2d \sin \theta$$
$$\text{thus, } n\lambda = 2d \sin \theta \qquad (4.23)$$

This is known as **Bragg's law** for X-ray diffraction.

What it says is that if we know the wavelength, λ, of the X-rays entering into the crystal, and if we can measure the angle θ of the diffracted X-rays coming out of the crystal, then we know the spacing (referred to as d-spacing) between the atomic planes.

$$d = n\lambda/2 \sin \theta$$

Again it is important to point out that this diffraction will occur only if the rays are in phase when they emerge, and only at the appropriate values of n (1, 2, 3, etc.) and λ.

In theory, we could reorient the crystal so that another atomic plane is exposed and measure the d-spacing between all atomic planes in the crystal, eventually leading us to determine the crystal structure and the size of the unit cell.

X-RAY DIFFRACTION METHODS

There are three different experimental methods for X-ray diffraction that we are going to look at: The Laue method, the rotating crystal method and the powder (Debye–Scherrer) method.

LAUE DIFFRACTION

Orientation of Single Crystals

Laue diffraction is most often used for mounting single crystals in a precisely known orientation. When the

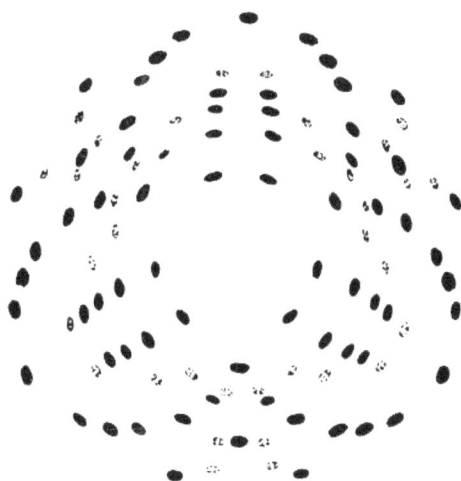

Figure 4.17 An example of a Laue photograph

incoming beam is parallel to a high-symmetry direction of the crystal, the Laue pattern also has high symmetry. In cubic crystals, an incoming beam parallel to one of the unit cell edges (a <001> direction) produces Laue patterns with 4-fold symmetry. An incoming beam parallel to the body diagonal of the unit cell produces a 3-fold symmetrical pattern of Laue spots (Figure 4.17).

Calculating Laue Angles

Figure 4.18 shows an xz-plane of a simple cubic crystal. Planes parallel to the y-direction are perpendicular to the plane of the drawing. A polychromatic X-ray beam enters the crystal along the *z*-direction [001]. The beam reflects off from the [201] planes, which are perpendicular to the [201] direction. Constructive interference of wavelengths that satisfy

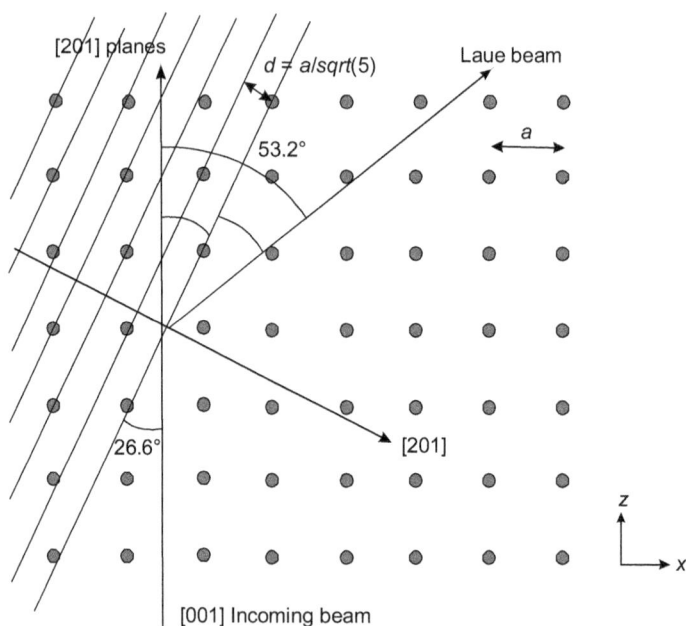

Figure 4.18   Laue Pattern

the Bragg's condition will give rise to the indicated Laue beam, which produces a spot on the film.

The direction of this beam is found by geometry—it does not depend on the lattice constant. The glancing angle of the incoming beam with the reflecting set of planes is given by the inner product of the normalized direction vectors.

$$\sin\theta = \frac{h\cdot h' + k\cdot k' + l\cdot l'}{\sqrt{h^2 + k^2 + l^2}\cdot\sqrt{h'^2 + k'^2 + l'^2}} \qquad (4.24)$$

As indicated in Figure 4.18, this gives that the angles of the diffracting [201] planes and the incoming and transmitted beam are both equal to 26.6°. Also the angle of reflection must be equal to the angle of incidence. Taken together, we find an angle of 53.2° between the Laue beam and the X-rays that go straight through the crystal.

The wavelength of a diffracted beam needs to satisfy the Bragg's condition. The glancing angle of incidence is given by equation 4.24, in our case 26.6°. The distance between the planes is calculated from the indices assigned to a Laue beam and the lattice parameter of the cubic crystal is given by

$$d = \frac{a}{\sqrt{h^2 + k^2 + l^2}} \qquad (4.25)$$

Method

The Laue method was the first diffraction method ever used, and it reproduces von Laue's original experiment. A beam of white radiation, the continuous spectrum from an X-ray tube, is allowed to fall on a fixed single crystal. The Bragg angle θ is therefore fixed for every set of planes in the crystal, and each set picks out and diffracts that particular wavelength which satisfies the Bragg's law for the particular values of d and θ involved. Each diffracted beam thus has a different wavelength.

There are two variations of the Laue method, depending on the relative positions of source, crystal, and film. In each, the film is flat and placed perpendicular to the incident beam. The film in the **transmission Laue method** (Figure 4.19a) is placed behind the crystal so as to record the beams diffracted in the forward direction. This method is so called because the diffracted beams are partially transmitted through the crystal. In the **back reflection Laue**

method (Figure 4.19b), the film is placed between the crystal and the X-ray source. The incident beam passing through a hole in the film and the beams diffracted in a backward direction are recorded.

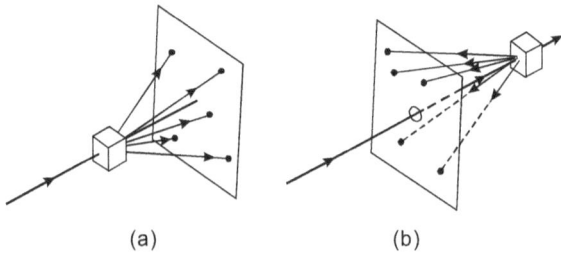

(a)   (b)

Figure 4.19   (a) Transmission and (b) Back reflection Laue methods

In either method, the diffracted beams form an array of spots on the film as shown in the Figure 4.20. This array of spots is commonly called a pattern, but the term is not used in any strict sense and does not imply any periodic arrangement of the spots. On the contrary, the spots are seen to lie on certain curves, as shown by the lines drawn

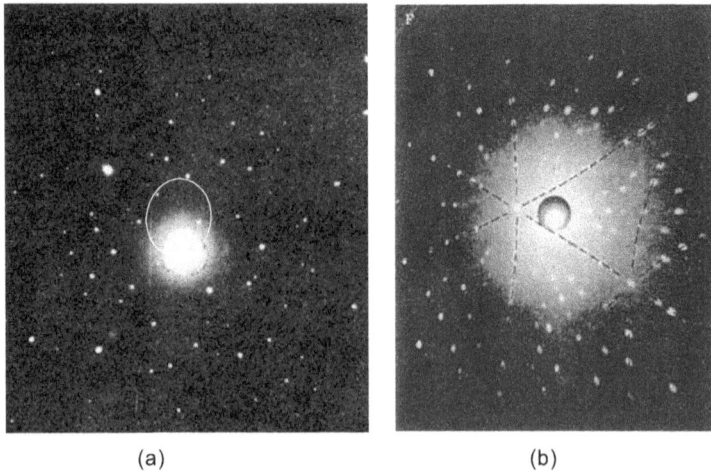

(a)   (b)

Figure 4.20   (a) Transmission and   (b) Back reflection patterns

on the photographs. These curves are generally ellipses or hyperbolas for transmission patterns (Figure 4.20a) and hyperbolas for back reflection patterns (Figure 4.20b).

The spots lying on any curve are reflections from planes belonging to one zone. This is due to the fact that the Laue reflections from planes of a zone all lie on the surface of an imaginary cone whose axis is the zone axis. As shown in Figure 4.21a, one side of the cone is tangent to the

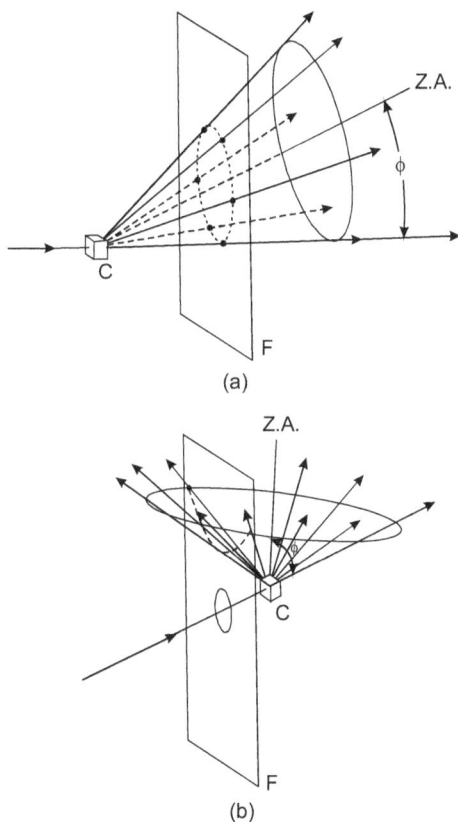

(a)

(b)

Figure 4.21   Location of Laue spots (a) on ellipses in transmission method and (b) on hyperbolas in back reflection method. (C—crystal, F—film, Z.A.—zone axis)

transmitted beam, and the angle of inclination ϕ of the zone axis to the transmitted beam is equal to the semiapex of the cone. A film placed as shown intersects the cone in an imaginary ellipse passing through the centre of the film, the diffraction spots from the planes of a zone being arranged on this ellipse. When the angle ϕ exceeds $45°$, a film placed between the crystal and the X-ray source to record the back reflection pattern will intersect the cone in a hyperbola as shown in Figure 4.21b.

The positions of the spots on the film, for both the transmission and the back reflection method, depend on the orientation of the crystal relative to the incident beam, and the spots themselves become distorted and smeared out if the crystal has been bent or twisted in anyway. These facts account for the two main uses of the Laue methods: the determination of crystal orientation and the assessment of crystal quality.

ROTATING CRYSTAL METHOD

In the rotating crystal method, a single crystal is mounted with one of its axes, or some important crystallographic direction, normal to a monochromatic X-ray beam. A cylindrical film is placed around it and the crystal is rotated about the chosen direction, the axis of the film coinciding with the axis of rotation of the crystal (Figure 4.22).

As the crystal rotates, a particular set of lattice planes will, for an instant, make the correct Bragg angle for reflection of the monochromatic incident beam, and at that instant a reflected beam will be formed. The reflected beams are again located on imaginary cones but now the cone axes coincide with the rotation axes. The result is that the spots on the film when the film is laid out flat, lie on imaginary horizontal lines. Since the crystal is rotated about only one axis, the Bragg angle does not take all possible values between 0° and 90° for every set of planes. Not every set therefore is able to produce a diffracted beam; sets perpendicular or almost perpendicular to the rotation axis are examples.

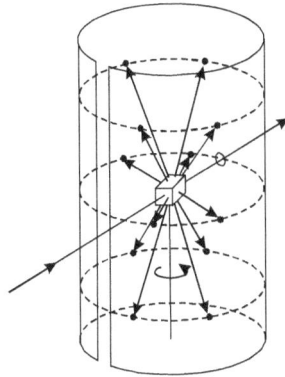

Figure 4.22   Rotating crystal method

X-RAY POWDER DIFFRACTION

X-ray powder diffraction is a non-destructive technique widely applied for the characterization of crystalline materials. This method has been traditionally used for phase identification, quantitative analysis and the determination of structure imperfections. Important advances in structural studies of materials ranging from high temperature superconductors and high-pressure research have relied heavily on the powder diffraction technique. Some solids can be prepared only as microcrystalline powders and hence their structures cannot be determined using single-crystal diffraction techniques. Also, the structures of some materials that are in the form of hydrocarbons and resins cannot be determined by single-crystal diffraction methods. In such cases we can determine the structure of the material using powder diffraction data. The ability to determine crystal structures using powder diffraction promises to open up many avenues in structural sciences. Single-crystal X-ray diffraction, although simple, is often very difficult to arrange in practice since it requires a single crystal to be grown. This may be difficult for some samples, and is impossible for multi-phase samples.

Although the diffraction of X-rays from powders (microcrystalline samples) had been discovered by Max von Laue shortly after his historical experiment on a single-crystal of KCl (which he crushed, on purpose, to verify his fresh hypothesis on the nature of such experiment), the powder diffraction (PD) technique has been seldom used as a structural tool. But some structural hypotheses on simple ionic compounds could be easily verified on the basis of experimental PD data, mostly collected by the pioneering Debye–Scherrer–Hull technique. It became very clear that PD traces could be used as fingerprints in qualitative analysis of crystalline materials of different origins.

Since the concept of molecular structure plays a central role in chemistry, much work has been done to structurally characterize as many molecules as possible. Surely, single-crystal X-ray diffraction is the easiest, quickest and the most definitive method applied. However, there are many compounds, which, for different reasons, do not afford single crystals of suitable size and quality. It would be unfortunate if the structure of a class of molecules, or even that of an individual compound, were ignored, simply because single crystals could not be easily obtained.

With the improvement of the experimental and numerical techniques now available, the determination of crystal structures from powder diffraction data has recently become a viable tool for assessing a number of structures, which could not be solved by more conventional (i.e., single-crystal) methods. However, the procedure that eventually leads to the correct structural model is by no means simple, and requires a lot of efforts, which have no real counterpart in the single-crystal technique. For example, while it is well known that poor single-crystal diffraction data can still afford gross molecular features which are often enough from an analytical point of view, the use of less-than-perfect powder diffraction data may lead nowhere, hampering, for example, the very basic procedures of finding the lattice constants (indexing) or phasing of the reflections. Some generating amorphous species, but seldom microcrystalline powder products can be found.

Principle of Powder Diffraction

From the Bragg's equation $n\lambda = 2d \sin \theta$, we can see that $\sin \theta$ is a measure of λ/d. We can choose the incident angle by rotating the crystal relative to the beam and the wavelength is fixed. Thus we obtain the interplanar spacing d. We can now conclude that any set of planes in a crystal will reflect an X-ray beam if the set of planes is at right angles ($\theta = 90°$) to the incident beam. But there arises another question whether the planes will reflect the beam strongly or not. The intensity of the reflected beam is proportional to the product of the intensity of the incident beam and the concentration of electrons in the reflecting plane. Thus if we know the unit cell dimensions and the atomic number of each of the atoms, we can calculate the concentration of electrons and hence the intensity of the reflected beam.

Now considering the reverse situation, if we know the size of the unit cell and the intensities of the reflections, we can calculate the positions of atoms and also the relative number of electrons per atom. It is obvious that all compounds with different formulae or unit cells have different collections of d-spacing and different intensities of reflections. The observed patterns of spacing and intensities can thus be used to identify an unknown compound in a specific crystalline phase.

If monochromatic X-ray radiation is taken instead of white light and the crystal is placed in front of the beam, there will be only one reflected beam for one particular angle of incidence. If the crystal is now rotated around the incident ray direction without changing the incident angle, the reflected beam will describe a cone with the crystal at the apex of the

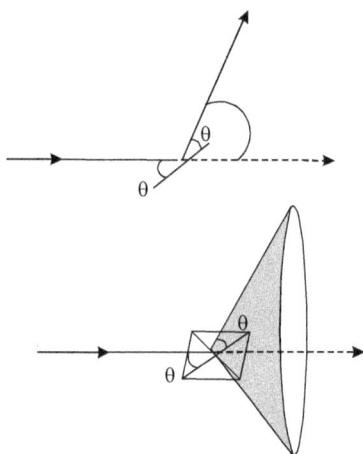

Figure 4.23   Formation of diffracted cone of radiation in the powder method

cone as shown in Figure 4.23. When there are hundreds of crystals, there are many reflected beams and when the crystals are rotated about their axes of incident X-ray beam, a series of cones are formed as shown in Figure 4.24a. If a powdered sample is placed in the path of X-rays, there will be a continuous series of point reflections lying along the arc of the cone. This is the basis of powder method that is used in X-ray crystallography to determine the unknown samples. For every set of crystal planes, one or more crystals will satisfy Bragg's condition. To understand the principles involved, consider a particular (*hkl*) reflection. One or more particles will be oriented so that their (*hkl*) planes obey Bragg's reflection condition. Figure 4.23 shows one plane in the set.

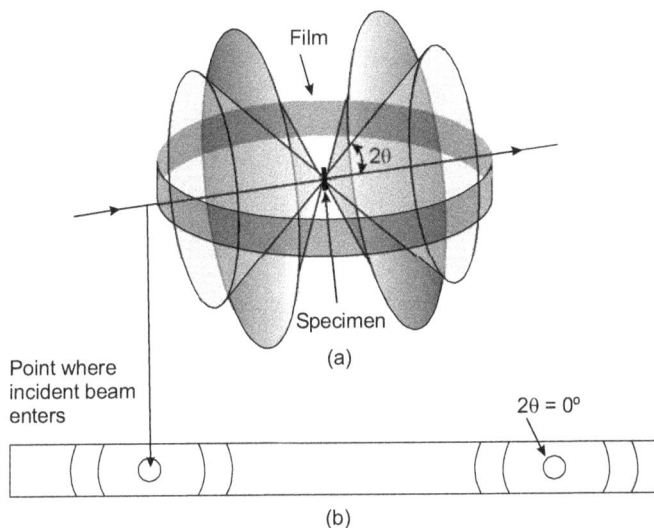

Figure 4.24   Debye–Scherrer powder method; (a) relation of film to specimen and incident beam (b) appearance of film when laid out flat

If the plane is now rotated such that the angle of incidence is kept constant, the reflected beam travels over the surface of the cone with the axis coinciding with the transmitted beam. Though this rotation does not occur in powder method, combined effect of some reflections from the (*hkl*) planes make the correct Bragg angle with the incident beam and thus have the form of a cone of diffracted radiation. Thus the (*hkl*) reflection from a powder sample produces many cones.

Methods of Powder Diffraction Pattern

The main methods of studying powders have led to the investigation of the atomic arrangements. The methods employed are the Debye–Scherrer method and diffractometry.

Debye–Scherrer method In Debye–Scherrer method, a narrow strip of film is curved in a short cylinder with the specimen placed on the axis and the incident beam is directed at right angles to the axis. The cones of diffracted radiation intersect the cylindrical strip in lines and when the strip is laid straight, the resulting pattern is as shown in Figure 4.24b. Each pattern is made up of small spots each from one particle and the spots are so close to each other that they appear as a continuous line. These lines are generally curved and when $2\theta = 90°$, they form a straight line. From the measured position of a given diffraction line, θ can be determined and if we know the wavelength λ, we can calculate the d-spacings of the lattice planes.

If the shape and size of the unit cell are known, the position of all the possible diffraction lines can be predicted. The Debye–Scherrer method is widely used especially in metallurgy. For this technique, the sample is contained in a holder and a strip of film with entry and exit holes for the X-rays is placed round the sample. Following exposure, the film is removed from the apparatus and flattened out. The resulting image consists of concentric dark arcs (Figure 4.25), which are the regions that have been exposed to X-rays.

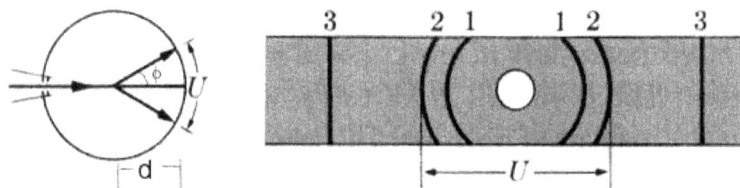

Figure 4.25 Appearance of the film

Bisecting the arcs on the film will give the positions of the entrance ($2\theta = 180°$) and the exit ($2\theta = 0°$) holes. By calculating the distance of the arcs from the holes, it is possible to obtain the diffraction angle. The diffractometer apparatus is commonly sized so that 1-mm distances on the film correspond very closely to 1° angle for convenience. Thus the diffraction peaks can be interpreted.

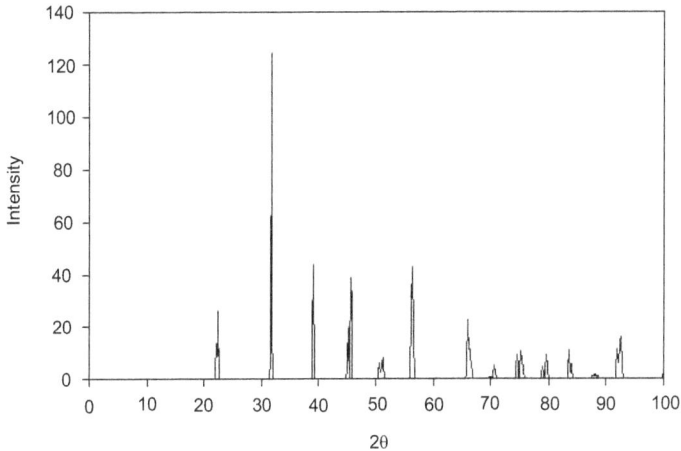

Figure 4.26 Powder diffraction pattern

A typical powder diffraction pattern for ($BaTiO_3$) samples is as shown in Figure 4.26.

The result from the film method for a sample of a perovskite structure crystal ($BaTiO_3$) is as shown in Figure 4.27.

Exit hole   Perovskite   Entrance hole

Figure 4.27 Photographic film

By combining the two results, it can be clearly seen that the two methods produce equivalent results, with both the angle of diffraction peak and relative intensity clearly seen in both cases (Figure 4.28).

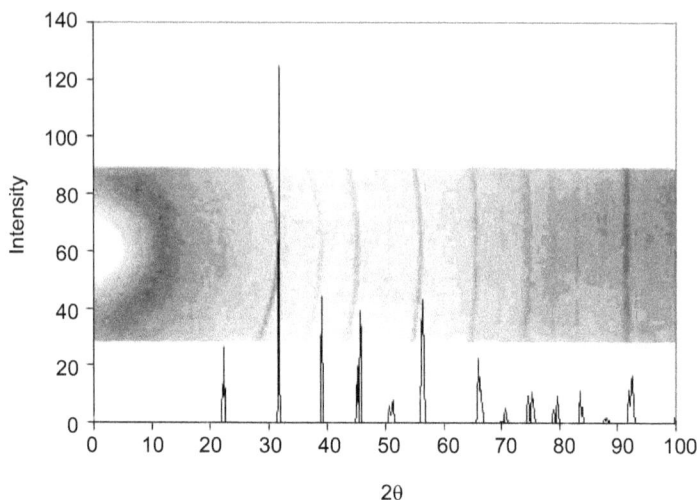

Figure 4.28 Powder and film method

Interpretation of Powder Photographs

The structure determination from the powder diffraction can be divided into three stages.

- Unit cell determination
- Structure solution
- Structure refinement

Unit cell determination The size and shape of the unit cell is determined from the positions of the lines. This is called indexing the pattern. Now let us consider that the sample is known to have cubic structure, but we do not know which cubic structure it has. After exposure of the sample to X-ray beam in the Debye–Scherrer method, the strip is removed and the positions are measured as follows:

The distance along the film from a diffraction line to the centre of the hole for the transmitted direct beam is measured. This is taken as S_1. This is shown in Figure 4.29.

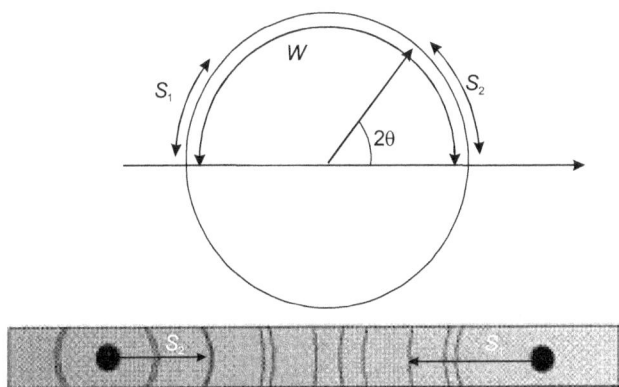

Figure 4.29 Interpretation of powder photograph

When $2\theta > 90°$, there is back reflection and S_2 is measured. This is the distance from the beam entry point. Now we know that S_1 corresponds to an angle 2θ. Also the distance between the holes W is obtained when $\theta = 180°$. Using this information we find that

$$\theta = \frac{\pi S_1}{2W}$$

$$\frac{\theta}{2} = \frac{\pi}{2}\left(1 - \frac{S_2}{W}\right)$$

Bragg's Law states that $n\lambda = 2d \sin \theta$. The interplanar spacing is given by

$$d_{hkl} = \frac{a}{\sqrt{h^2 + k^2 + l^2}}$$

where, a is the lattice parameter. This gives

$$\sin^2 \theta = \frac{\lambda^2}{4a^2}(h^2 + k^2 + l^2) \qquad [\because 2d \sin \theta = n\lambda, \text{ where } n = 1]$$

Now from the arcs obtained in the powder method and the diffraction method we have the values of S_1, θ and $\sin^2 \theta$. If we take all the diffraction lines into consideration, then the values of $\sin^2 \theta$ should form a pattern related to the values of (hkl). Multiplying the values of $\sin^2 \theta$ by a constant (K) gives nearly an integer value for each of the $h^2 + k^2 + l^2$ values. This can be interpreted from Table 4.1 in which the measurements are taken for a sample.

Table 4.1   Measurements obtained from the sample

				Now assign whole numbers		
$\lambda = 1.54$ Å		$W = 180$ mm				
S_1 (mm)	θ	$\sin^2 \theta$	$K \sin^2 \theta$	$h^2 + k^2 + l^2$	hkl	a(Å)
38	19.0	0.11	3.0	3	111	4.05
45	22.5	0.15	4.1	4	200	4.02
66	33.0	0.30	8.2	8	220	4.02
78	39.0	0.40	10.9	11	311	4.04
83	41.5	0.45	12.3	12	222	4.02
97	49.5	0.58	15.8	16	400	4.04
113	56.5	0.70	19.1	19	331	4.03
118	59.0	0.73	19.9	20	420	4.04
139	69.5	0.88	24.0	24	422	4.01
168	84.0	0.99	27.0	27	511	4.03

Multiply by 27.3 in this case

Now the integer values of $h^2 + k^2 + l^2$ are equated with the integer values of hkl and the corresponding measurements are obtained. This is shown in Table 4.2 and the type of cubic system is identified as simple cubic.

Sometimes for some structures, some of the arcs may be missing as in the case of bcc and fcc structures. Then it becomes easy to identify the structure. The value of the lattice parameter can also be calculated as from the 2θ positions of the hkl lines and may be refined for systematic errors.

Table 4.2 Values of $h^2 + k^2 + l^2$ and *hkl*

Possible planes in any structure:

$h^2 + k^2 + l^2$	*hkl*
1	100
2	110
3	111
4	200
5	210
6	211
8	220
9	221
10	310
11	311
12	222
13	320
14	321
16	400

Notice that 7 and 15 are missing.

Thus, in the case of cubic crystals, knowing the values of $(h^2 + k^2 + l^2)$ for different planes, one can easily identify the type of cubic crystals in the following manner.

Crystal systems	Values of $(h^2 + k^2 + l^2)$
Simple cubic	1, 2, 3, 4, 5, 6, 8, 9, 10, 11, 12, 13, 14, 16
Body centred cubic	2, 4, 6, 8, 10, 12, 14, 16
Face centred cubic	3, 4, 8, 11, 12, 16
Diamond cubic	3, 8, 11, 16

Similarly if the individual values of *hkl* sets (i.e., 311 or 220) are purely odd or even, immediately one can identify that the given type of crystal belongs to fcc lattice.

Structure solution In structure solution, an initial approximate structure is obtained from experimental data without having any prior knowledge of the arrangement of atoms and molecules. This is a very important phase in the determination of structure. There are two techniques for structure solution.

- Traditional approach
- Direct space approach

Traditional approach   In traditional approach, the intensities are taken from the powder pattern and they are used in the calculation that is used for single-crystal diffraction data. But this may not be a reliable approach as there are many overlapping peaks and hence the intensity values obtained are not exact. This problem can be overcome by using improved techniques for extracting intensities or new strategies in which the pattern is obtained without extracting intensities. The experimental pattern generates trial structures. This comparison is done using an appropriate *R*-factor. Most of the direct space approaches use the weighted *R*-factor, which is R_{wp}. This is given by

$$R_{wp} = 100 \times \frac{\sum W_i(y_i(\text{obs}) - y_i(\text{calc}))^2}{W_i(y_i(\text{obs}))^2}$$

Here $y_i(\text{obs})$ is the intensity of the *i*th data point in the experimental powder pattern. $y_i(\text{calc})$ is the intensity of the *i*th data point in the calculated powder diffraction profile. W_i is the weighting factor for the *i*th data point. R_{wp} thus considers the intensities point by point instead of the integrated intensities. This reduces the peak overlap and uses the powder diffraction data as measured.

Direct space approach   The basis of direct space strategy is to find a hypersurface $R(\Gamma)$ to find the global minimum. Here Γ represents the set of variables that define the surface. These variables define the position, orientation and the intramolecular geometry of each molecule. The position is defined by $\{x, y, z\}$ orientation by $\{\theta, \phi, \psi\}$ and the intramolecular geometry is represented by the set of variables $\{\tau 1, \tau 2, \tau 3, \dots \tau n\}$. The intensities, peak positions and the peak shapes can be determined prior to the structure solution method by using fitting procedure. There are many techniques for determining the lowest value on the $R(\Gamma)$ surface. They are

   i. Monte Carlo method

  ii. Simulated annealing

 iii. Genetic algorithm

Applications

X-ray powder diffraction has opened up new avenues in the studies of structures. It has a number of applications.

- *Qualitative analysis* From the pattern, the d-spacings are recorded and the relative intensities of the 10 strongest lines are measured and are compared with the patterns of the known compounds. This comparison is done with the help of a powder diffraction file that contains the patterns of some standard compounds divided into subdivisions— minerals, inorganic and organic.

- *Quantitative analysis* For a two-component mixture, the relative concentration of each of the components can be obtained by measuring the relative intensities of the strong non-overlapping lines, each belonging to the two components.

- *Structure of alloys* An alloy is a mixture of two or more elements. If the composition is uniform, it produces a typical powder diffraction pattern. If one of the components precipitates, it produces separate lines on the powder pattern corresponding to the component.

- *Stress determination in metals* If there is a stress in a metal, then the angle of the diffraction cone changes because of a change in the *d*-spacing due to stress. By measuring the changes in the cone angle, accurate measurements of stress can be made. In addition, stress invariably broadens diffraction peaks unless it is absolutely uniform on an atomic scale.

- *Determination of particle size* As the size of the crystallite decreases, the angular spread of the reflection increases. The half height width can be used as a measure of the mean particle size of the sample.

- *Identification and raw material evaluation* For some complex materials, it is difficult to analyse the pattern. But since similar materials exhibit similar patterns, for example, we can determine the structure of different clays as a cement material by comparing with acceptable clay and thus relate structure to properties.

Limitations

- The individual peak intensities are difficult to obtain because in powder diffraction, a 3-D pattern is reduced to a 1-D pattern and analysis is done. This leads to both accidental and exact peak overlap.

- The symmetry of crystals cannot be obtained accurately in powder diffraction pattern.

- Preferred orientation can lead to inaccurate peak intensities. But both rotating the sample about its normal and rocking it about each data point can overcome this.

CRYSTAL STRUCTURE DETERMINATION

SCATTERING FACTOR

When an X-ray beam encounters an atom, each electron in it scatters part of the radiation coherently in accordance with the Thomson equation. One might also expect the nucleus to take part in the coherent scattering, since it also bears a charge and should be capable of oscillating under the influence of the incident beam. However, *the nucleus has an extremely large mass relative to that of the electron and cannot be made to oscillate to any appreciable event*; in fact the Thomson equation says that the intensity of coherent scattering is inversely proportional to the square of the mass of the scattering particle. The net effect is that *coherent scattering by an atom is only due to the electrons contained in that atom.*

An atom of atomic number Z, i.e., an atom containing Z electrons, scatters a wave whose amplitude is Z times the amplitude of the wave scattered by a single electron, provided the scattering is in the forward direction ($2\theta = 0$), because the waves scattered by all the electrons of the atom are then in phase and the amplitudes of all the scattered waves can be added directly.

This is not true for other directions of scattering. The fact that the electrons of an atom are situated at different points in space introduces differences in phase between the waves scattered by different electrons. Consider in Figure 4.30 in which the electrons are shown as points arranged around the central nucleus. The waves scattered in the forward direction by electrons A and B are exactly in phase on a wavefront such as XX', because each wave has travelled the same distance before and after scattering. The other scattered waves shown in Figure 4.30 however, have a path difference equal to (CB − AD) and are thus somewhat out

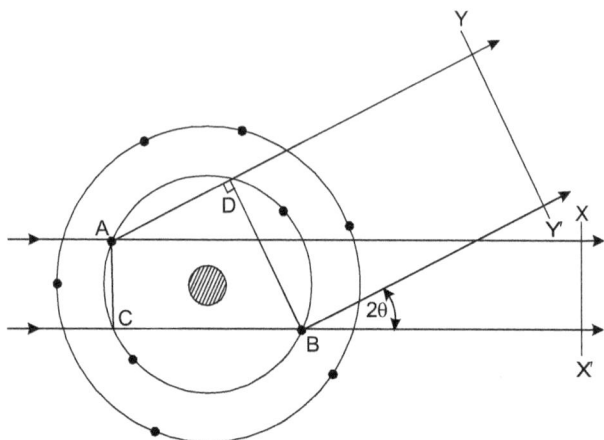

Figure 4.30   X-ray scattering by an atom

of phase along a wavefront such as YY', the path difference being less than one wavelength. Partial interference occurs between the waves scattered by A and B with the result that the net amplitude of the wave scattered in this direction is less than that of the wave scattered by the same electrons in the forward direction.

A quantity f, the **atomic scattering factor**, is used to describe the "efficiency" of scattering of a given atom in a given direction. It is defined as a ratio of amplitudes:

$$f = \frac{\text{amplitude of the wave scattered by an atom}}{\text{amplitude of the wave scattered by an electron}}$$

From what has been said already, it is clear that $f = Z$ for any atom scattering in the forward direction. As θ increases, however, the waves scattered by individual electrons become more and more out of phase and f decreases. The atomic scattering factor depends also on the wavelength of the incident beam; at a fixed value of θ, f will be smaller since the path differences will be larger relative to the wavelength, leading to greater interferences between the scattered beams. The actual calculation of f involves $\sin\theta$ rather than θ, so that the net effect is that f decreases as the quantity ($\sin\theta/\lambda$) increases. The scattering factor f is sometimes called the **form factor**, because it depends on the way in which electrons are distributed around the nucleus.

The curve showing the typical variation of f, in this for copper, is given in Figure 4.31. Note that the curve begins at the atomic number of copper ($Z = 29$), and decreases to very low values for scattering in the backward direction (θ near 90°) or for very short wavelengths. Since the intensity of a wave is proportional to the square of its amplitude, a curve of scattered intensity from an atom can be obtained simply by squaring the ordinates of a curve such as in the Figure 4.31.

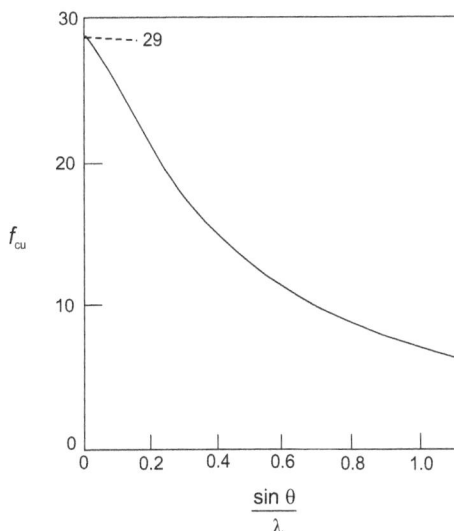

Figure 4.31 The atomic scattering factor of copper

The scattering just discussed, whose amplitude is expressed in terms of the atomic scattering factor, is coherent, or unmodified scattering, which is the only kind capable of being diffracted. On the other hand, incoherent, or Compton-modified scattering occurs at the same time. Since the latter is due to collisions of quanta with loosely bound electrons, its intensity relative to that of the unmodified radiation increases as the proportion of loosely bound electrons increases. The intensity of Compton-modified radiation thus increases as the atomic number Z decreases. It is for this reason that it is difficult to obtain good diffraction photographs of organic materials, which contain light elements such as carbon, oxygen, and hydrogen, since the strong Compton-modified scattering from these substances darkens the background of the photograph and makes it difficult to see the diffraction lines formed by the unmodified radiation. It is also found that the intensity of the modified radiation increases as the quantity $(\sin \theta)/\lambda$ increases.

STRUCTURE FACTOR

Depending upon the symmetry, there may be more than one molecule in the unit cell. The contribution to the scattering by the unit cell contents (different atoms each having its own scattering factor f_j, position at (x_j, y_j, z_j) the fractional coordinates of the jth atom) is termed as **structure factor**, characterized by a quantity F_{hkl}. This is the complex quantity

$$F_{hkl} = |F_{hkl}| e^{(i\phi_{hkl})} \qquad (4.26)$$

whose *magnitude is the amplitude of the scattered wave and whose direction in the complex plane is determined by the phase of the scattered wave.* The measured intensities from diffraction experiment are proportional to the square of their respective structure factor amplitudes.

$$I_{hkl} \propto |F_{hkl}|^2 \qquad (4.27)$$

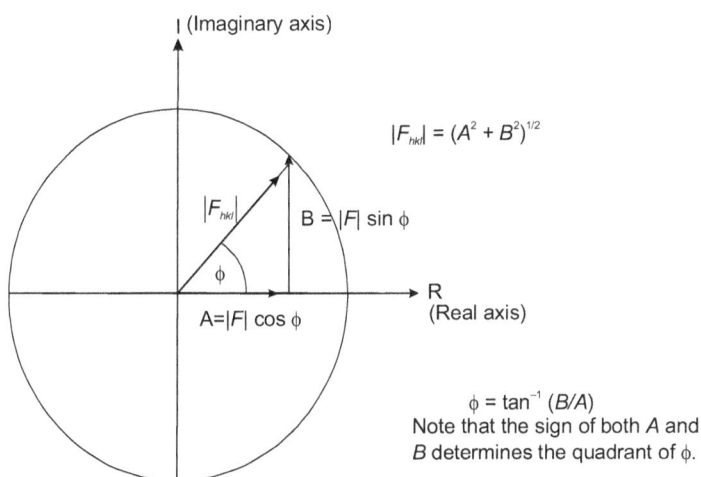

$|F_{hkl}| = (A^2 + B^2)^{1/2}$

$B = |F| \sin \phi$

$|F_{hkl}|$

ϕ

$A = |F| \cos \phi$

R (Real axis)

I (Imaginary axis)

$\phi = \tan^{-1} (B/A)$
Note that the sign of both *A* and
B determines the quadrant of ϕ.

Figure 4.32 Vector representation of *F* in complex plane

Since the factor F_{hkl} depends on the arrangement of matter in the specific crystal under discussion, i.e., on its crystal structure, the factor is generally called structure factor. Since the scattering unit in any molecule is the electron, the structure factor eventually depends on the summation over the general distribution of the electrons with respect to the centre of the atoms. The representation of the complex quantity F_{hkl} is shown in Figure 4.32. For a summation over j = 1 to N atoms of the unit cell, the structure factor of a Bragg's reflection can be written as

$$F_{hkl} = f_1 e^{i\phi_1} + f_2 e^{i\phi_2} + f_3 e^{i\phi_3} + ... \qquad (4.28)$$

$$= \sum_{j=1}^{N} f_j e^{i\phi_j} \qquad (4.29)$$

Each group of terms in this condensed structure factor (equation 4.28) represents the scattering by an atom in the unit cell (Figure 4.33).

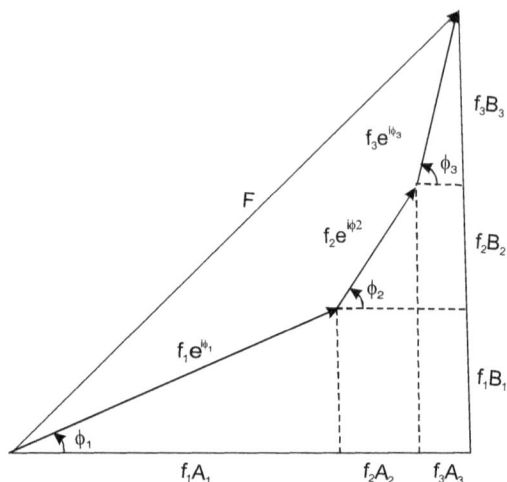

Figure 4.33   Condensed structure factor (F)

The phase angle ϕ_j can be expressed as a function of fractional coordinates x_j, y_j and z_j of the atoms of the unit cell.

$$\phi_j = 2\pi(hx_j + ky_j + lz_j) \qquad (4.30)$$

Making use of equation 4.30 in 4.28, we obtain an expression for the structure factor as

$$F_{hkl} = f_1 e^{i2\pi(hx_1 + ky_1 + lz_1)} + f_2 e^{i2\pi(hx_2 + ky_2 + lz_2)} +$$
$$f_3 e^{i2\pi(hx_3 + ky_3 + lz_3)} + \dots \qquad (4.31)$$

$$= \sum_j f_j e^{i2\pi(hx_j + ky_j + lz_j)} \qquad (4.32)$$

From equation 4.32 it is clear that F_{hkl} is a complex quantity

$$F_{hkl} = \left| F_{hkl} \right| e^{i\phi_{hkl}} \qquad (4.33)$$

$$F_{hkl} = |F_{hkl}|(\cos\phi_{hkl} + i\sin\phi_{hkl}) = A_{hkl} + iB_{hkl} \qquad (4.34)$$

where, $A_{hkl} = |F_{hkl}|\cos\phi_{hkl}$; $B_{hkl} = |F_{hkl}|\sin\phi_{hkl}$ from which

$$\tan\phi_{hkl} = \frac{B_{hkl}}{A_{hkl}} \qquad (4.35)$$

where, ϕ_{hkl} is called the phase of the Bragg's reflection *hkl*.

$$\phi_{hkl} = \tan^{-1}\left[\frac{B_{hkl}}{A_{hkl}}\right] \qquad (4.36)$$

Since A_{hkl} and B_{hkl} can take any values, ϕ_{hkl} in general can assume any values between 0 to 2π. From equation 4.32,

$$F_{hkl} = \sum_{j=1}^{N} f_j \cos 2\pi(hx_j + ky_j + lz_j)$$
$$+ i\sum_{j=1}^{N} f_j \sin 2\pi(hx_j + ky_j + lz_j) \qquad (4.37)$$

$$F_{hkl} = A_{hkl} + iB_{hkl} \qquad (4.38)$$

Thus, $A_{hkl} = \sum_{j=1}^{N} f_j \cos 2\pi(hx_j + ky_j + lz_j)$

and

$$B_{hkl} = \sum_{j=1}^{N} f_j \sin 2\pi(hx_j + ky_j + lz_j) \qquad (4.39)$$

As F_{hkl} is a complex quantity,

$$|F_{hkl}| = \sqrt{(A_{hkl}^2 + B_{hkl}^2)} \qquad (4.40)$$

and this quantity is called the calculated structure factor amplitude ($|F_c|$) which is a function of the coordinates of the atoms in the unit cell. To distinguish this quantity from the observed structure factor amplitude, $|F_0|$ ($|F_0| \propto \sqrt{I_{hkl}}$), which is obtainable from experimental measurement of Bragg intensities, $|F_{hkl}|$ as defined in equation 4.40 is usually called $|F_c|$, denoting calculated structure factor magnitudes of a Bragg reflection *hkl*. (In $|F_0|$ and $|F_c|$ one usually suppresses the reflection indices *h*, *k*, *l*.)

CENTROSYMMETRIC CRYSTAL AND THE PHASE PROBLEM

In this case, for every atom *j* in the unit cell situated at the fractional coordinates (x_j, y_j, z_j), there is an identical atom in the unit cell at ($-x_j$, $-y_j$, $-z_j$). Thus out of a total number of *N* atoms in the unit cell, N/2 atoms will be at fractional coordinates (x_j, y_j, z_j) and the remaining N/2 atoms will be at the fractional coordinates ($-x_j$, $-y_j$, $-z_j$). In this case, the structure factor F_{hkl} (equation 4.37) becomes

$$F_{hkl} = \sum_{j=1}^{N/2} f_j e^{2\pi i [h(x_j) + k(y_j) + l(z_j)]} + \sum_{j=N/2}^{N} f_j e^{2\pi i [h(-x_j) + k(-y_j) + l(-z_j)]} \qquad (4.41)$$

Making use of the trigonometric relation, $e^{i\theta} = \cos\theta + i\sin\theta$ and $e^{-i\theta} = \cos\theta - i\sin\theta$, the equation 4.41 can be simplified as

$$F_{hkl} = 2\sum_{j=1}^{N/2} f_j \cos 2\pi (hx_j + ky_j + lz_j) \qquad (4.42)$$

$\therefore F_{hkl} = \sum_{j=1}^{N} f_j \cos 2\pi (hx_j + ky_j + lz_j)$ which is simply equal to A_{hkl} (equation 4.39), i.e., in centrosymmetric case, $B_{hkl} = 0$. Referring to equation 4.35, $\tan \phi_{hkl} = 0$ and ϕ_{hkl} has only two choices namely 0 or π in the case of centrosymmetric crystal when compared to any value between 0 to 2π in the case of non-centrosymmetric crystals. Thus the phase problem gets simplified in centrosymmetric cases.

$$F_{hkl} = |F_{hkl}| e^{i\phi_{hkl}},$$

$$\phi_{hkl} = 0 \text{ or } \pi.$$

When, $\phi_{hkl} = 0$, $F_{hkl} = |F_{hkl}| e^{i0} = |F_{hkl}|$ and

when, $\phi_{hkl} = \pi$, $F_{hkl} = |F_{hkl}| e^{i\pi} = |F_{hkl}|(\cos \pi + i \sin \pi)$

$$= -|F_{hkl}| \qquad (4.43)$$

Thus the phase problem in the case of centrosymmetric crystals gets simplified with the assignment of algebraic signs (+ or −) to the structure factor magnitudes obtained from experiment.

NEED FOR PHASE

The ultimate aim in the three-dimensional structure elucidation of any molecule is to obtain the three-dimensional fractional coordinates (x_j, y_j, z_j) of j atoms constituting the unit cell. Since atoms contain electrons, the location of atomic site is equivalent to finding the centre of electron cloud for each of the atom. To locate an atom, the meaningful way is to compute the electron densities at various points in the unit cell; the electron density maxima correspond to the atomic site. The electron density maxima x, y, z in the unit cell can be written as

$$\rho_{x,y,z} = \frac{1}{V} \sum_{h} \sum_{k} \sum_{l}^{\infty} F_{hkl} \, e^{-2\pi i (hx+ky+lz)} \qquad (4.44)$$

where, V is the volume of the unit cell (which can be calculated by unit cell parameters a, b, c, α, β and γ).

The triple summation is over the three miller indices *hkl* varying from $-\infty$ to ∞.

In order to locate the electron density maxima from the left hand side of the equation 4.44, one should be able to compute the right hand side of the equation 4.44 in which F_{hkl} is the coefficient. (This equation is also called a Fourier synthesis or electron density equation). Thus if we know the structure factors (inverse space from diffraction by electrons) we can calculate the actual real structure (the density of the electrons in real space).

Figure 4.34 shows the electron density derived from actual data. The electron density is calculated for each point (x, y, z) in space and points of equal value are connected forming the characteristic wire grid presentation.

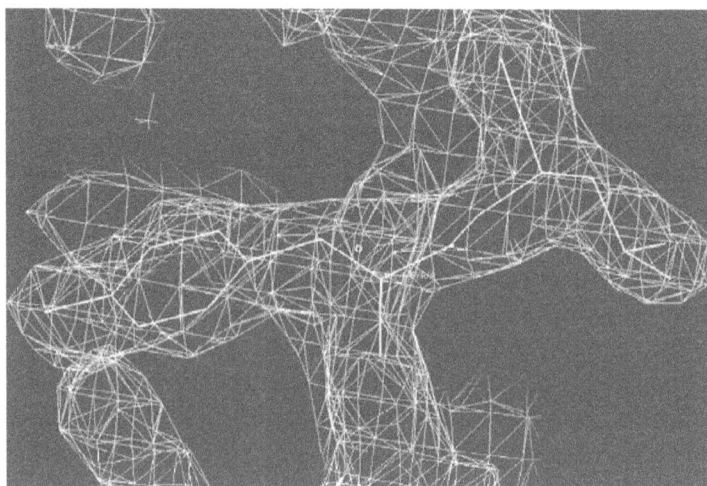

Figure 4.34   Electron density

Since F_{hkl} involves two components, $|F_{hkl}|$ and ϕ_{hkl} ($F_{hkl} = |F_{hkl}| e^{i\phi_{hkl}}$), the computation of electron density is possible only if both the structure factor magnitudes and phases of the Bragg's reflections are known. But unfortunately experimental measurement gives only structure factor amplitudes $|F_{hkl}|$ and thus ϕ_{hkl} remains unknown. This bottleneck is known as **phase problem** in crystallography namely the non-availability of phases of Bragg's reflections, which are required along with the known structure factor magnitudes, to locate the electron density maxima in order to eventually locate an atomic site. Thus unless the phases ϕ_{hkl} are known, atomic sites could not be located.

Methods of Solving the Phase Problem

There are various methods to solve the phase problem, which will be discussed in a nutshell. Some of them are

   i. Patterson method

   ii. Isomorphous replacement method

   iii. Anomalous dispersion method

   iv. Direct methods

These methods can be used effectively to locate the approximate positions of the atoms of the trial structure of a molecule in the unit cell.

Patterson Method

Patterson introduced a function in 1934, which is a Fourier transform of the set of squared but not phased reflection amplitudes ($h\ k\ l\ |F^2|$). This function does not produce an electron density map of the contents of the unit cell but rather a density map of the vectors between scattering objects in the cell. Because the densities in the Patterson map go as squares of the numbers of electrons of the scattering atoms, Patterson maps of crystals that contain heavy atoms are dominated by the vectors between heavy atoms. However, because the number of peaks in a Patterson map is also related to the square of the number of atoms, protein Patterson maps are rarely interpretable. If we carry out an inverse Fourier transform of the intensities (amplitudes squared), which would require only the measured data, the resulting map, is now called a **Patterson function** or **Patterson map**.

The Patterson function gives us a map of the vectors between atoms. In other words, if there is a peak of electron density for atom 1 at position $x_1(x, y, z)$ and a peak of electron density for atom 2 at position $x_2(-x, -y, -z)$, then the Patterson map will have peaks at positions given by x_2-x_1 and x_1-x_2.

Patterson synthesis is given by

$$P(u,v,w) = \sum_{hkl} I_{hkl} \cos 2\pi(hu + kv + lw) \tag{4.45}$$

where,

$u = 2x,$

$v = 2y$ and

$w = 2z.$

From this, (x, y, z) can be calculated as $(u/2, v/2, w/2)$.

The height of the peak in the Patterson map is proportional to the product of the heights of the two peaks in the electron density map. Figure 4.35 illustrates a Patterson

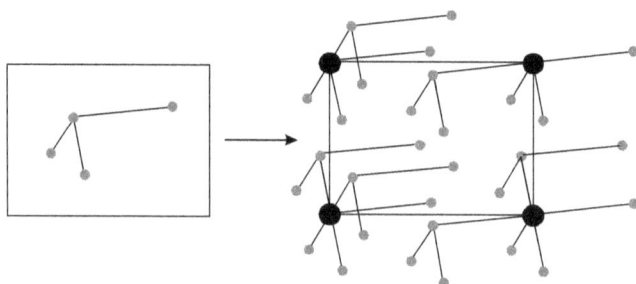

Figure 4.35   Patterson map

map corresponding to a cell with one molecule. It demonstrates that one can think of a Patterson as being a sum of images of the molecule, with each atom placed in turn on the origin.

For relatively small number of atoms, it is possible to work out the original positions of the atoms that would give rise to the observed Patterson peaks. This is called deconvoluting the Patterson. But it quickly becomes impossible to deconvolute a Patterson for larger molecules. If we have N atoms in a unit cell and the resolution of the data is high enough, there will be N separate electron density peaks in an electron density map. In a Patterson map, each of these N atoms has a vector to all N atoms, so that there are N^2 vectors. N of these will be self-vectors from an atom to itself, which will accumulate as a big origin peak, but that still leaves $N^2 - N$ non-origin peaks to sort out. If N is a small number, say 10, then we will have a larger but feasible number of non-origin Patterson peaks to deal with (90 for N = 10). But if N were 1000, which would be more in the range seen for protein crystals, then there would be 999,000 non-origin Patterson peaks. The difference between the electron density map and Patterson map is shown in Figure 4.36.

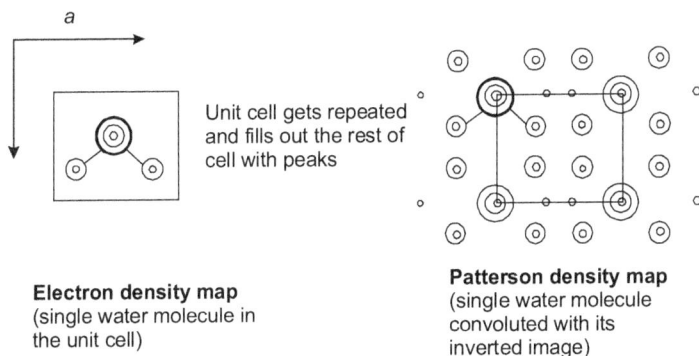

Electron density map
(single water molecule in
the unit cell)

Patterson density map
(single water molecule
convoluted with its
inverted image)

Figure 4.36   Electron density map and Patterson map

In 1954, Perutz and co-workers calculated a **difference Patterson** $(|F_{PH}| - |F_P|)^2$ with the amplitudes of a mercury-labelled haemoglobin crystal (derivative) and the amplitudes of an isomorphous native haemoglobin crystal. Here, the scattering of the light atoms is mathematically removed (leaving noise, of course) so that the difference Patterson map ought to show simply the vectors between heavy atoms.

Isomorphous Replacement

Isomorphous replacement is the first phasing technique, which became available to macromolecular crystallographers. Let us assume we could replace one atom in the crystal (usually a water atom) with a heavy metal atom (the big black one) in Figure 4.37, without altering the structure. Such a procedure, called isomorphous replacement, is actually possible by soaking a protein crystal, which consists to about 50% of solvent, in a heavy metal solution.

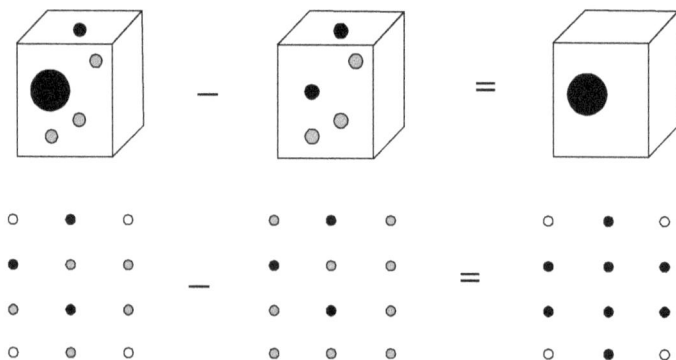

Figure 4.37   Isomorphous replacement

Heavy atoms such as transition metals, lanthanides, uranium and even noble gases under pressure, can quite *successfully be soaked into crystals through solvent channels and frequently bind to well defined sites in the native protein*. Such a protein crystal is then called a heavy atom derivative (crystal). Based on the assumption that the derivative crystal retains its structure (i.e., in fact isomorphous), we guess that we should be able to derive some information on the structure factor amplitudes for the heavy metal from the differences between the derivative and the native dataset.

We can now imagine creating in real space a **difference crystal** by subtracting the native crystal from the derivative crystal.

Apparently, we have reduced the problem to a much simpler one: in real space, imaginary difference crystal would contain only the heavy atom (minus some contributions from whatever it has replaced). The same scenario is conceivable in reciprocal space with a diffraction pattern. The resulting difference pattern contains still as many reflections, but they originate from the heavy atom only in the crystal structure. Note that a prerequisite for this subtraction of patterns is that they remain isomorphous, i.e., we do not change the structural terms in the structure factor summation.

Finding the heavy atoms The benefit of a diffraction pattern created by a single atom (or a few) is that we can actually find its positions, and therefore the phases. We still cannot use a Fourier synthesis to find the electron density of the heavy atoms, because we still lack their phases. But we can create a phase-less Patterson synthesis, based on the intensities (not the square root) of the reflections. This map (Figure 4.38) contains peaks at the **distance vectors** (u, v, w) between atoms in the structure. The peak height is proportional to the mean square of the electrons (Z) of the atoms located at the ends of the vector (this is why we prefer heavy atoms like Hg, Pt, Au for soaking) and the number of peaks $P = N^2 - N$ (N is the number of atoms in the structure). P becomes large very rapidly, and for ease of interpretation, we want only few atoms in the derivative.

Some sophisticated methods can actually solve quite large numbers of atoms, even up to hundreds. Once heavy atom positions of two derivatives are determined, phasing equations can be solved and initial protein phases are determined.

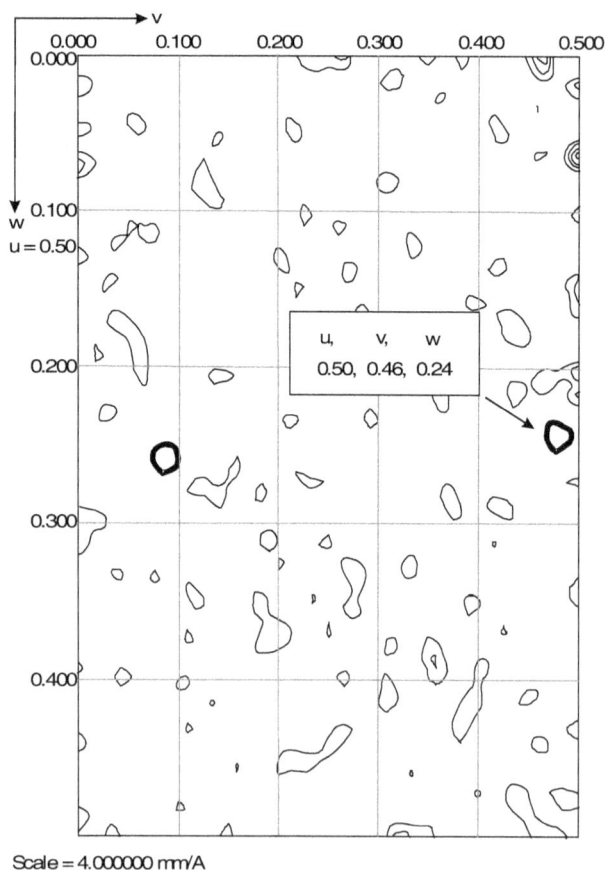

Figure 4.38   Isomorphous difference Patterson map

It is based on the fact that the structure factor F_{hkl} for a certain reflection hkl is a simple summation of all individual atomic scattering contributions.

$$F_{hkl} = \sum_{j=1}^{N} f_j \exp\left[2\pi i(hx_j + ky_j + lz_j)\right] \qquad (4.46)$$

This allows us to do simple arithmetic with F such as $F_{PH} = F_P + F_H$. Let us assume that F_P is the structure factor of a reflection for the protein, and F_{PH} for the same protein now including a heavy atom. Unfortunately, in the absence of phase information we cannot apply the simple subtraction $F_H = F_{PH} - F_P$ and end up with the structure factors (and thus positions) of the heavy atom in one step, at least not in the generic non-centrosymmetric case of interest to macromolecular crystallographers. Note also that $|F_H| = |F_{PH}| - |F_P|$ is not correct! (Figure 4.39). This is obvious from the fact that a negative difference in the intensities may occur—only implying that the phase of F_H is somehow opposite to F_P. We need to separate phase terms from magnitudes in order to do math correctly with a mixture of structure factors F and structure amplitudes $|F|$.

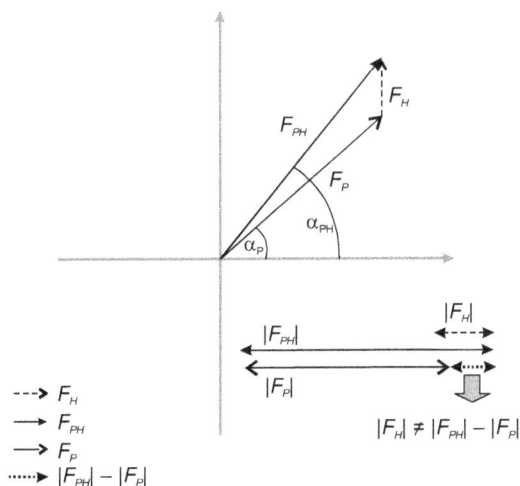

Figure 4.39 Relationship between native and derivative structure factors

From the experiment, we know only the magnitudes $|F_{PH}|$ (derivative) and $|F_P|$ (protein) which can be represented in the complex plane as a circle of radius $|F_{PH}|$ and $|F_P|$, respectively. If we know both the magnitude and the phase of F_H we can draw both circles offset by vector F_H (dotted arrows) and obtain two solutions for possible phase values for F_P (Figure 4.40).

The phase and magnitude of F_H can be calculated easily if we know the positions of a heavy metal. At this point, it is clear that the best phase we can obtain from the two solutions is the mean in between the two possibilities, and the phase error can be quite large. In real cases, F_H is much shorter (---> type of vector) and the phase error will be quite large, somewhat below 90°. We realize that on average, **SIR (single isomorphous replacement)** phases will

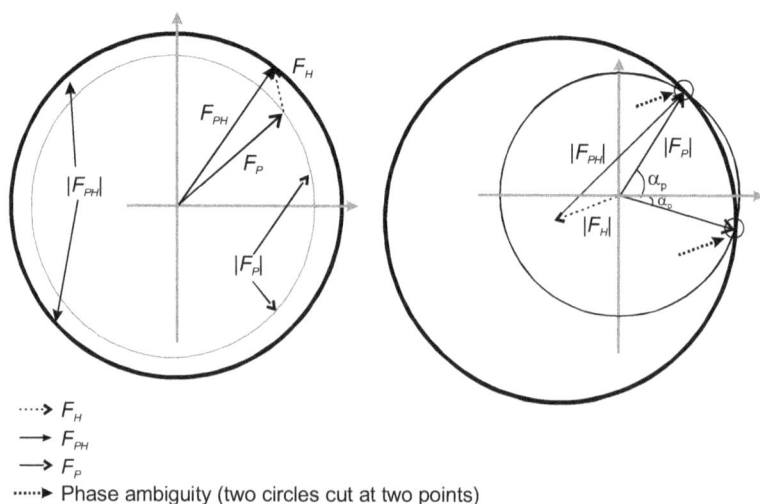

····> F_H
⟶ F_{PH}
⟶ F_P
······► Phase ambiguity (two circles cut at two points)

Figure 4.40   Phase determination in SIR

be better with larger contributions from the heavy metal. In order to eliminate the **phase ambiguity**, we can prepare a **second derivative** and repeat the procedure. Provided the heavy atom is not binding at the same position, we can now obtain a unique solution for $\alpha(p)$, the phase of F_p(Figure 4.41).

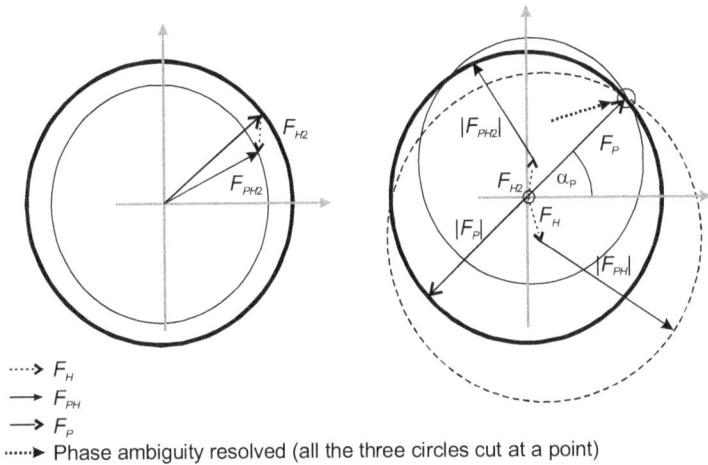

····> F_H
⟶ F_{PH}
⟶ F_P
·····► Phase ambiguity resolved (all the three circles cut at a point)

Figure 4.41 Phase determination in MIR

We have now, at least in theory, an exact solution for the phase angle of F_p. The theory is based on two assumptions: a) ideal isomorphism and b) exact heavy atom positions, neither of which are perfectly met, for practical and experimental reasons in the first case and for theoretical reasons in the second. In Figure 4.41, it means that the phasing circles may not intersect exactly in one spot, and another derivative may be necessary to improve the quality of the phases. The method is therefore called **multiple isomorphous replacement (MIR)**.

Anomalous Dispersion

Most electrons in the atoms that make up a crystal will interact identically with X-rays. If placed at the origin of the crystal, they will diffract with a relative phase of zero. Because of this, pairs of diffraction spots obey Friedel's law, which is illustrated in Figure 4.42. The thick black arrows indicate a diffraction event from the top of the planes. The atoms contribute to the diffraction pattern with phases determined by their relative distances from the planes, as indicated by the dashed arrows on Figure 4.42b. The thin arrows on Figure 4.42a indicate a very similar diffraction event, but from the bottom of the same planes. The angles of incidence and reflection are the same, and all that is different is, which side of the planes we're looking at. If the thick black arrows define planes with Miller indices ($h\,k\,l$), the same planes are defined from the other side with Miller indices ($\bar{h}\,\bar{k}\,\bar{l}$). The reflection with indices ($\bar{h}\,\bar{k}\,\bar{l}$) is referred to as the **Friedel mate** of ($h\,k\,l$). Atoms will contribute with the same phase shift, but where the phase shifts were positive, they would now be negative. This is illustrated in Figure 4.42b with the thin black arrows on the bottom, each of which has the opposite phase of the dashed arrows on the top. The effect of reversing the phases is to reflect the picture across the horizontal axis.

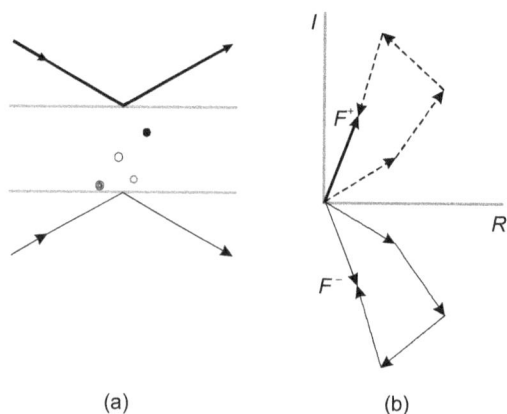

(a)   (b)

Figure 4.42 Normal scattering ($I_{hkl} = I_{\bar{h}\bar{k}\bar{l}}$); ($|F_{hkl}| = |F|_{\bar{h}\bar{k}\bar{l}}$)

When the X-ray photon energy is close to transition energy (such transitions are used, in fact, to generate X-rays with a characteristic wavelength and we often use a particular transition of electrons in copper), there will be a small shift in both the amplitude and phase of the scattered X-ray. This shift in amplitude and the phase is called **anomalous scattering**.

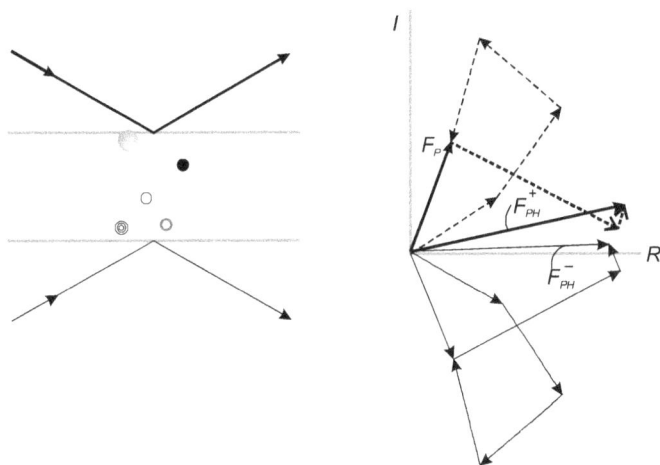

Figure 4.43   Violation of Friedel's law $(I_{hkl} \neq I_{\bar{h}\bar{k}\bar{l}})$; $(|F_{hkl}| \neq |F|_{\bar{h}\bar{k}\bar{l}})$

The phase shift in anomalous scattering leads to a breakdown of Friedel's law, as illustrated in Figure 4.43. Now we have added a heavy atom with an anomalous scattering component. It is convenient to represent the phase shift by adding a vector at 90 degrees to the normal scattering for the heavy atom. Significantly, this vector is at +90 degrees from the contribution of the anomalous scattering, regardless of which of the two Friedel mates we are looking at and this causes the symmetry to break down.

The effect is easier to see (and to use) if we take the Friedel mate and reverse the sign of its phase, i.e., reflect it across the horizontal axis (Figure 4.44). (Thinking of the structure factor as a complex number, this means that we reverse the sign of the imaginary component, the result of which is called the complex conjugate, indicated with an asterisk.)

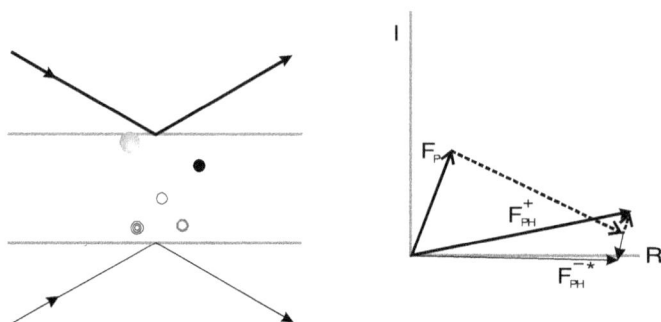

Figure 4.44 Anomalous scattering

Now we can see that the effect of anomalous scattering has been to make the amplitudes of the Friedel mates different. If we have a model for the anomalous scatterers in the crystal, we can draw vectors for their contribution to the structure factors for the Friedel mates and construct a Harker diagram, as in the case of MIR.

The anomalous scattering effect depends on the frequency of oscillation being similar to the natural frequency of the atom. So it is clear that the strength of the anomalous scattering effect depends on the wavelength of the X-rays, which will change both the normal scattering and the out-of-phase scattering of the anomalous scatterers.

In the presence of anomalous scattering the atomic scattering factor becomes a complex quantity defined as

$$f = f_0 + f' + if'' \quad (f' \text{ is usually negative})$$

By collecting data at several wavelengths near the absorption edge of an element in the crystal, we can obtain phase information analogous to that obtained from MIR. This technique is called **multiple-wavelength anomalous**

dispersion (MAD). One popular way to use MAD is to introduce selenomethionine in place of methionine residues in a protein while it is expressed. The selenium atoms (which replace the sulphur atoms) have a strong anomalous signal at wavelengths that can be obtained from synchrotron X-ray sources. MIR and MAD techniques are useful in the determination of crystal structures of macromolecules.

Direct Methods

Direct methods are used to calculate the phases by simple mathematical procedures from a single set of X-ray intensities. The three important basic assumptions on which the direct methods was established are

- positivity
- atomicity and
- randomness

In the initial stage, atoms are assumed to be of point atom type [as fall-off of intensity at high scattering angles is due to atomic size and atomic vibrations, $|F_{bkl}|$ is now replaced by $|E_{bkl}|$ which do not vary with $\sin \theta /\lambda$].

Direct method procedures

1. Normalized structure factors $|E_{bkl}|$ is given by

$$\left|E_{bkl}\right|^2 = \frac{\left|F_{bkl}\right|^2}{\varepsilon \sum_j^N f_j^2}$$

where, ε is a constant, (already calculated).

Only high $|E_{bkl}|$ values (those greater than 1.5 are used) signify greater validity of the probabilistic estimate. The magnitudes $|E_{bkl}|$ of the normalized structure factors are uniquely determined by crystal structure and are independent of the choice of origin but the values of phases ϕ_{bkl} depend on the choice of the origin. But there exists certain linear combinations of phases, which are called **structure invariants**, whose values are determined by the structure alone and are independent of the choice of origin. In shifting the origin by a vector, the phases turn out to be origin-dependent while the amplitudes are not. But there are certain specific phase relations of the form given below which do not change with shift in origin. Let us see the phase determination in detail.

Phase determination in direct methods   Direct methods are the mathematical methods or probabilistic methods, which arrive at the phases directly from the measured Bragg's intensities. Linear combinations of phase relationships are used in direct methods. Direct methods suppose that the real crystal with continuous electron density $\rho(\vec{r})$ is replaced by an idealized one, the unit cell of which consists of N discrete, non-vibrating point atoms. Then the structure factor $F_{\vec{b}}$ is replaced by the normalized structure factor $E_{\vec{b}}$ (structure factor corresponding to point atoms) and is defined by

$$E_{\vec{b}} = \left| E_{\vec{b}} \right| \exp(i\phi_{\vec{b}}) \tag{4.47}$$

$$E_{\vec{b}} = \frac{1}{\sigma_2^{1/2}} \sum_{j=1}^{N} f_j \exp(2\pi i \vec{b} \cdot \vec{r}_j) \tag{4.48}$$

$$\left\langle E_{\vec{b}} \exp(-2\pi i \vec{b}.\vec{r}) \right\rangle_{\vec{b}} = \frac{1}{\sigma_2^{1/2}} \left\langle \sum_{j=1}^{N} f_j \exp(2\pi i \vec{b}.(\vec{r}_j - \vec{r})) \right\rangle_{\vec{b}} \tag{4.49}$$

$$= \frac{f_j}{\sigma_2^{1/2}} \quad (\text{if } \vec{r} = \vec{r}_j) \tag{4.50}$$

$$= 0 \quad (\text{if } \vec{r} \neq \vec{r}_j) \tag{4.51}$$

where,

f_j is the zero-angle atomic scattering factor,

$\vec{r_j}$ is the position vector of the atom labelled j, and

$$\sigma_2 = \sum_{j=1}^{N} f_j^{\,2} \qquad (4.52)$$

From the equations 4.49, 4.50, 4.51 one sees that only at atomic positions, non-vanishing terms (equation 4.51) exist for the synthesis of the L.H.S. of equation 4.50.

In the X-ray diffraction case, the f_j s are equal to the atomic numbers Z_j and are therefore all-positive; in the neutron diffraction case, some of the f_j may be negative. It is noteworthy that recent advances are equally valid whether or not some of the f_j are negative, so that the application to neutron diffraction is automatic.

The magnitudes $\left| E_{\vec{b}} \right|$ are related to the experimentally measured Bragg intensities by

$$\left| E_{\vec{b}} \right|^2 = \frac{\left| F_{\vec{b}} \right|^2}{\varepsilon \sum f_j^{\,2}} \qquad (4.53)$$

Thus, eventhough $\left| E_{\vec{b}} \right|$ of the normalized structure factor is available from the observed magnitude $\left| F_{\vec{b}} \right|$, at least approximately, the phases defined in equation 4.47 cannot be determined experimentally. Equation 4.50 implies only the elucidation of the 3N components of the N position vectors $\vec{r_j}$ rather than the much more complicated electron density function $\rho(r)$. Equation 4.49, after equating its real and imaginary parts, is a system of 2n equations where n is the number of known magnitudes $\left| E_{\vec{b}} \right|$. The unknown parameters are the n phases of $\phi_{\vec{h}}$ and 3N components of $\vec{r}_j$, and hence $n + 3N$ unknowns, in total. Since $2n > n + 3N$, the phase problem when reformulated in the above way is solvable in principle.

Before dealing with the phase relationships which are useful in direct methods to obtain the individual phases, one has to look into the effect of changing the origin on the normalized structure factors. Let $\vec{r}_j$ be the position vector of the jth atom at P with respect to the first origin and let this origin be shifted to another point whose position vector with respect to initial origin is $\overrightarrow{\Delta r}$ (Figure 4.45).

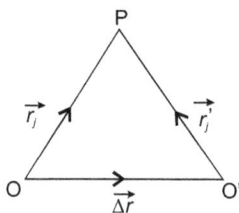

Figure 4.45   Effect of origin shift

Now the position of the point in the new origin is related to the old position vector by

$$\vec{r}_j = \overrightarrow{\Delta r} + \vec{r}_j'$$

$$\vec{r}_j' = \vec{r}_j - \overrightarrow{\Delta r} \tag{4.54}$$

and now $E_{\vec{h}}$ is replaced by $E_{\vec{h}}'$ given by

$$E_{\vec{h}}' = \frac{1}{\sigma_2^{1/2}} \sum_{j=1}^{N} f_j \exp(2\pi i \vec{h} \cdot \vec{r}_j') \tag{4.55}$$

$$E_{\vec{h}}' = \frac{1}{\sigma_2^{1/2}} \sum_{j=1}^{N} f_j \exp(2\pi i \vec{h} \cdot (\vec{r}_j - \overrightarrow{\Delta r})) \tag{4.56}$$

$$E_{\vec{b}}' = \frac{\exp(-2\pi i \vec{b} \cdot \overrightarrow{\Delta r})}{\sigma_2^{1/2}} \sum_{j=1}^{N} f_j \exp(2\pi i \vec{b} \cdot \vec{r_j}) \qquad (4.57)$$

i.e., $E_{\vec{b}}' = E_{\vec{b}} \exp(-2\pi i \vec{b} \cdot \overrightarrow{\Delta r})$

Using equation 4.47, the equation above can be rewritten as

$$\left|E_{\vec{b}}'\right| e^{i\phi_{\vec{b}}'} = \left|E_{\vec{b}}\right| e^{i\phi_{\vec{b}}} \; e^{(-2\pi i \vec{b} \cdot \overrightarrow{\Delta r})}$$

$$\text{i.e.,} \quad \left|E_{\vec{b}}'\right| = \left|E_{\vec{b}}\right|$$

$$\phi_{\vec{b}}' = \phi_{\vec{b}} - 2\pi(\vec{b} \cdot \overrightarrow{\Delta r})$$

For another reflection $\vec{k}$, one obtains a similar relation,

$$\phi_{\vec{k}}' = \phi_{\vec{k}} - 2\pi(\vec{k} \cdot \overrightarrow{\Delta r})$$

For one more reflection $\vec{l}$, one has the same relation

$$\phi_{\vec{l}}' = \phi_{\vec{l}} - 2\pi(\vec{l} \cdot \overrightarrow{\Delta r})$$

The above relationship clearly indicates the phase changes with respect to a shift in origin. Adding the above three-phase relations, one obtains,

$$\phi_{\vec{b}}' + \phi_{\vec{k}}' + \phi_{\vec{l}}' = \phi_{\vec{b}} + \phi_{\vec{k}} + \phi_{\vec{l}} - 2\pi(\vec{b} + \vec{k} + \vec{l}) \cdot \overrightarrow{\Delta r} \qquad (4.58)$$

$\varphi_{\vec{b}}'$ is the phase of structure factor $E_{\vec{b}}'$ with respect to new origin. $\vec{b}$ is any reflection defined by the Miller indices, say $b_1 k_1 l_1$; $\vec{k}$ is another reflection, say, $b_2 k_2 l_2$; $\vec{l}$ is one more reflection, say, $(b_3 k_3 l_3)$.

The above three-phase relation equation 4.58 also clearly indicates that the combination of these phases also changes with respect to the shift in origin. But when $\vec{b} + \vec{k} + \vec{l} = 0$, (which is possible because $b_1 k_1 l_1$, $b_2 k_2 l_2$, $b_3 k_3 l_3$ can take any values, positive or negative integers including zero;

$$\vec{b} + \vec{k} + \vec{l} \equiv (b_1 + b_2 + b_3, \ k_1 + k_2 + k_3, \ l_1 + l_2 + l_3) \equiv (0,0,0)$$

$$\phi'_{\vec{b}} + \phi'_{\vec{k}} + \phi'_{\vec{l}} = \phi_{\vec{b}} + \phi_{\vec{k}} + \phi_{\vec{l}} = \phi$$

indicating that the linear combination of three phases of reflection vectors $\vec{b}, \vec{k}, \vec{l}$ such that $\vec{b} + \vec{k} + \vec{l} = 0$, *does not change with respect to shift in origin*. In other words, the sum of the three phases $\phi_{\vec{b}} + \phi_{\vec{k}} + \phi_{\vec{l}}$ is a structure invariant, a three-phase structure invariant provided $\vec{b} + \vec{k} + \vec{l} = 0$. This is called **triplet relation**.

The above argument can be extended for one more reflection vector $\vec{m}$ subject to the condition that $\vec{b} + \vec{k} + \vec{l} + \vec{m} = 0$. Now we have a four-phase relation $\phi'_{\vec{b}} + \phi'_{\vec{k}} + \phi'_{\vec{l}} + \phi'_{\vec{m}} = \phi_{\vec{b}} + \phi_{\vec{k}} + \phi_{\vec{l}} + \phi_{\vec{m}}$ called a **quartet relationship** (with either four magnitudes with main terms or with seven magnitudes with three more cross terms).

Probability relationships have been worked out in the literature (Hauptman, 1975a, b; 1976b) for triplets, quartets, quintets and sextets.

In the presence of symmetry elements, the origin cannot be chosen arbitrarily and it has to lie on the symmetry axis. This way, due to space group symmetries, the choices of origin become restricted and we have permissible origins only. The phase relations which remain unchanged or invariant when the origin is shifted to within those permitted by space group symmetry are called **structure semi-invariants**.

The structure invariants and semi-invariants have been tabulated for all the space groups (Hauptman and Karle,

1953, 1956, 1959; Karle and Hauptman, 1961; Lessinger and Wondratschek, 1975).

To each of the reciprocal vectors $\vec{b}, \vec{k}, \vec{l}$ there are three corresponding normalized structure factor magnitudes $\left|E_{\vec{b}}\right|, \left|E_{\vec{k}}\right|, \left|E_{\vec{l}}\right|$ and three phases $\phi_{\vec{b}}, \phi_{\vec{k}}, \phi_{\vec{l}}$. Denoting $\phi = \phi_{\vec{b}} + \phi_{\vec{k}} + \phi_{\vec{l}}$, the conditional probability distribution of ϕ, with the given three magnitudes, is given by (Cochran, 1955; Hauptman, 1976a, b)

$$P_{1/3} = P\left(\phi\,|, |E_{\vec{b}}|, |E_{\vec{k}}|, |E_{\vec{l}}|\right)$$

$$\approx \frac{1}{2\pi I_0(A_{bk})} e^{A_{bk}\cos\phi}$$

where, $A_{bk} = \frac{2}{\sqrt{N}}\left|E_{\vec{b}}\, E_{\vec{k}}\, E_{\vec{l}}\right|$, for equal atom structures and I_0 is the modified Bessel function of order zero.

Figures 4.46 to 4.49 show typical conditional probability distribution functions.

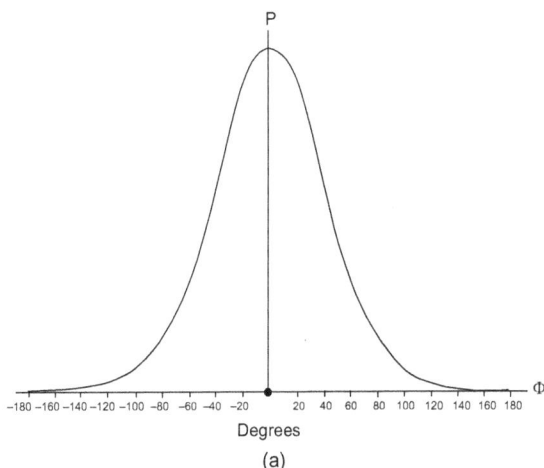

Figure 4.46 a) The distribution $P_{1/3}$ for $A = 2.316$ (*Continues*)

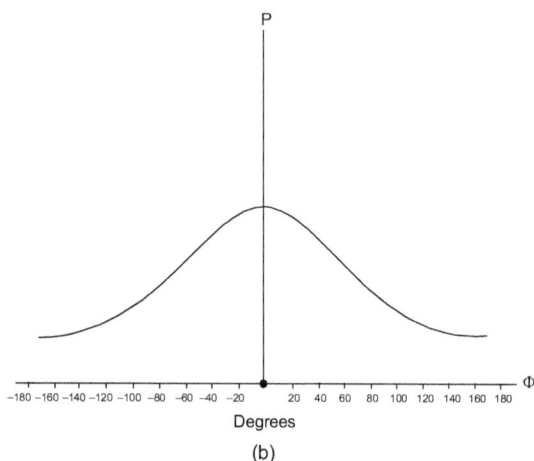

Figure 4.46   b) The distribution $P_{1/3}$ for A = 0.731

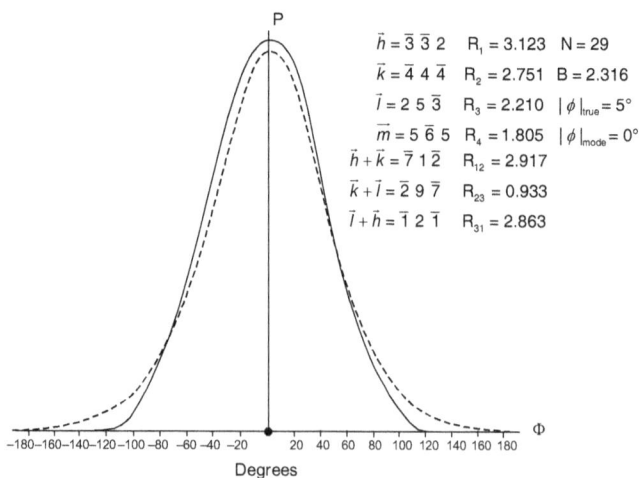

$\bar{h} = \bar{3}\ \bar{3}\ 2$ $R_1 = 3.123$ N = 29
$\bar{k} = \bar{4}\ 4\ \bar{4}$ $R_2 = 2.751$ B = 2.316
$\bar{l} = 2\ 5\ \bar{3}$ $R_3 = 2.210$ $|\phi|_{true} = 5°$
$\bar{m} = 5\ \bar{6}\ 5$ $R_4 = 1.805$ $|\phi|_{mode} = 0°$
$\bar{h} + \bar{k} = \bar{7}\ 1\ \bar{2}$ $R_{12} = 2.917$
$\bar{k} + \bar{l} = \bar{2}\ 9\ \bar{7}$ $R_{23} = 0.933$
$\bar{l} + \bar{h} = \bar{1}\ 2\ \bar{1}$ $R_{31} = 2.863$

Figure 4.47   The conditional probability distribution of the quartet with 7 magnitudes (solid line) and for quartet with 4 magnitudes (dashed line) The mode of both is zero always.

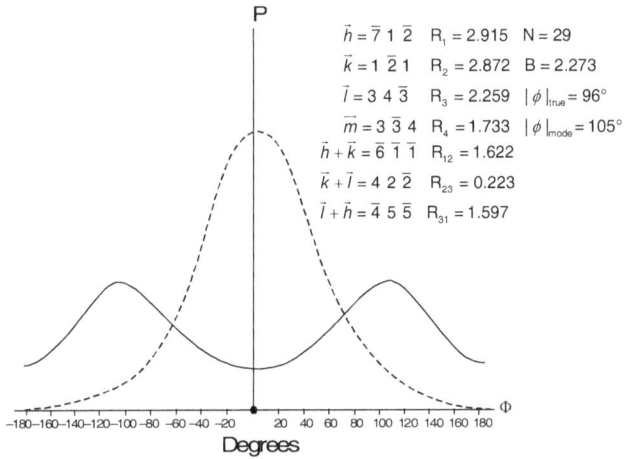

P

$\vec{h} = \bar{7}\,1\,\bar{2}$ $R_1 = 2.915$ $N = 29$

$\vec{k} = 1\,\bar{2}\,1$ $R_2 = 2.872$ $B = 2.273$

$\vec{l} = 3\,4\,\bar{3}$ $R_3 = 2.259$ $|\phi|_{true} = 96°$

$\overline{m} = 3\,\bar{3}\,4$ $R_4 = 1.733$ $|\phi|_{mode} = 105°$

$\vec{h} + \vec{k} = \bar{6}\,\bar{1}\,\bar{1}$ $R_{12} = 1.622$

$\vec{k} + \vec{l} = 4\,2\,\bar{2}$ $R_{23} = 0.223$

$\vec{l} + \vec{h} = \bar{4}\,5\,\bar{5}$ $R_{31} = 1.597$

Figure 4.48   The conditional probability distribution of the quartet with 7 magnitudes (solid line) and for quartet with 4 magnitudes (dashed line). The mode for 7 magnitudes is 105° and is zero for 4 magnitudes.

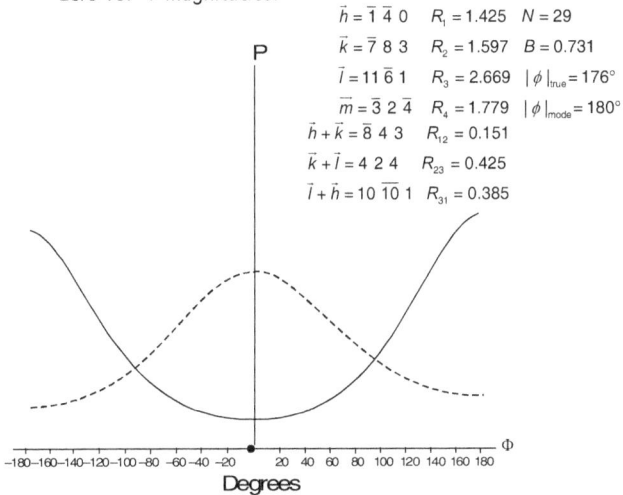

P

$\vec{h} = \bar{1}\,\bar{4}\,0$ $R_1 = 1.425$ $N = 29$

$\vec{k} = \bar{7}\,8\,3$ $R_2 = 1.597$ $B = 0.731$

$\vec{l} = 11\,\bar{6}\,1$ $R_3 = 2.669$ $|\phi|_{true} = 176°$

$\overline{m} = \bar{3}\,2\,\bar{4}$ $R_4 = 1.779$ $|\phi|_{mode} = 180°$

$\vec{h} + \vec{k} = \bar{8}\,4\,3$ $R_{12} = 0.151$

$\vec{k} + \vec{l} = 4\,2\,4$ $R_{23} = 0.425$

$\vec{l} + \vec{h} = 10\,\overline{10}\,1$ $R_{31} = 0.385$

Figure 4.49   The conditional probability distribution of the quartet with 7 magnitudes (solid line) and for quartet with 4 magnitudes (dashed line). The mode for 7 magnitudes is 180° and is zero for 4 magnitudes.

2. Convergence procedure is used to select a small set of reflections (usually four reflections) the starting set, and assigning their phases to be known. Only strong $|E_{hkl}|$ values are chosen in the generation of invariants so that the reliability of the probabilistic estimates has a direct relation to the normalized structure factor magnitudes entering in the probability formulae for **triplets** or **quartets**. Sets of three Bragg's reflections are selected with indices that satisfy the triple-product sign relationship (Σ_2 formula).

$$\phi(\vec{h}) \approx \left\langle \phi(\vec{k}) + \phi(\vec{h} - \vec{k}) \right\rangle_{\vec{k}}$$

where $\phi(\vec{h})$ is the phase angle of the Bragg's reflection *hkl*.

3. The phases of the above small set of reflections are now assumed to be known and knowing the values of two phases and the sum of the three phases (from the probability formulae), the value of the unknown phase can be found. This is called the phase propagation or phase extension. This is also termed as **divergence procedure**. This propagation and phase refinement are carried out using **tangent formula**

$$\tan\theta_h = \frac{\sum\limits_{k_j}\left|E_{k_j} E_{h-k_j}\right|\sin(\phi_{k_j} + \phi_{h-k_j})}{\sum\limits_{k_j}\left|E_{k_j} E_{h-k_j}\right|\cos(\phi_{k_j} + \phi_{h-k_j})} \qquad (4.59)$$

e.g. when $\vec{h} + \vec{k} + \vec{l} = 0$, $\phi_{\vec{h}} + \phi_{\vec{k}} + \phi_{\vec{l}} \approx 0$ (the triplet phase sum is probably equal to zero. Let $\vec{h}$ and $\vec{k}$ reflections be in the start set with phases say 0 and π, respectively. Using the above relation we find that $0 + \pi + \phi_{\vec{l}} \approx 0$. Hence $\phi_{\vec{l}} = \pi$ (phase of new reflection is now found).

4. Depending on the choices of the phase values for the reflections chosen for the origin and enantiomorph, the direct methods procedure become multisolution in nature from which the correct solution can be picked up. Before doing an E-map, the set with the lowest value for the combined figure of merit is selected as the correct one (in SHELXS 97 program) and the E-map is plotted for this set only.

    CFOM = R_α + {0 or (NQUAL-w$_n$)}, whichever is larger

    $$R_\alpha = \Sigma w \ (\alpha - \alpha_{est})^2 / \Sigma w \ (\alpha_{est})^2$$

    where, the weight w is $1/(\alpha_{est} + 5)$ (to avoid the largest αs dominating).

5. Since $\left|F_{\vec{b}}\right|$ and $\phi_{\vec{b}}$ are now known, an E-map could now be calculated for this set. This map will reveal mostly the entire structure. In case of incompleteness, the existing model is refined for few cycles (i.e., isotropic refinement) and a difference Fourier (a Fourier map whose coefficient is $\left|F_o\right| - \left|F_c\right|, \left|F_c\right|$ is the calculated structure factor amplitude using atoms in the incomplete model) will reveal the rest of the atoms.

Structure Refinement

The refinement of a structure begins with the trial structure obtained from the structure solution (say, from SHELXS 97 output) in order to get the correct atomic positions and the associated thermal parameters. The position and thermal parameters are adjusted to get a better fit between $|F_o|$ and $|F_c|$. Generally, a full-matrix least-squares refinement method is employed in small molecular structure refinement using the program SHELXL 97 (Sheldrick, 1997). The least-squares refinement method uses the squares of the differences between the observed and calculated structure factor amplitudes as a measure of their disagreement, and adjust the parameters so that the total disagreement is a minimum. The refinement is based on F_o^2 because it is impossible to refine on F using all the data which would involve taking the square root of a negative number for reflections with negative F_o^2 (i.e., background higher than the peak as a result of statistical foundation). The refinement of F_o^2 using all the data provides a good result for weakly diffracting crystals and in particular for pseudosymmetry problems.

The residual factor or reliability index defining the correctness of the model is given by

$$R_1 = \frac{\sum_{hkl} \left\| F_o \right| - \left| F_c \right\|}{\sum_{hkl} \left| F_o \right|} \qquad (4.60)$$

where,

$\left| F_o \right|$ = observed structure factor amplitude and

$\left| F_c \right|$ = calculated structure factor amplitude.

(This is sometimes denoted as R(F) also).

The summation is taken over all the observed reflections. *R*-value should be a minimum for the accurate model. A suitable weighting scheme is applied at the end of refinement procedure and the weighted *R*-factor is given by

$$wR_2 = \frac{\sum w_i \left(\left| F_o \right|^2 - \left| F_c \right|^2 \right)^2}{\sum w_i \left(\left| F_o \right|^2 \right)^2} \qquad (4.61)$$

METHODS OF RECORDING X-RAY DIFFRACTION

THE OSCILLATION METHOD

The oscillation method of recording the X-ray diffraction from a single crystal was apparently first practiced by de Broglie, Wagner and Seeman in the half-dozen years immediately following the discovery of X-ray diffraction by crystals. In this method, a large single crystal was rotated about a fixed axis set normal to the X-ray beam. The X-ray diffraction by the crystal was received and recorded (in those early days) on a flat photographic plate, also placed normal to the X-ray beam. In this way, the X-ray reflections directed along the generators of the Laue cones within the range of the photographic plate were recorded as seen in the Figure 4.50. The pattern we see corresponds to planes in reciprocal (diffraction) space slicing through the Ewald sphere so that only a limited amount of each lattice plane is in diffraction condition within the oscillation range. Entire datasets are built up by collecting contiguous series of such

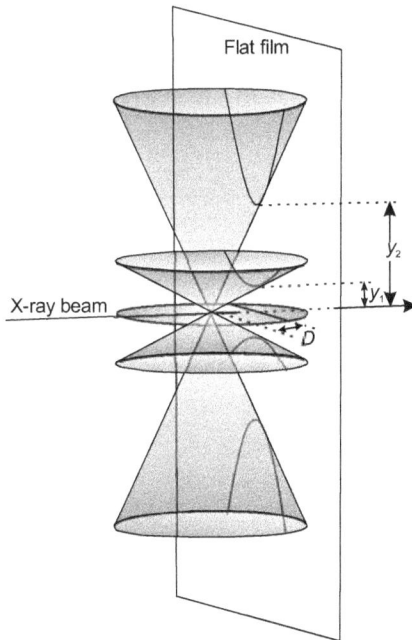

Figure 4.50 Recording of the Laue cones parallel to the rotation axis on a flat film

images to form a solid volume of rotation. This is known as the "rotation method" or oscillation method. The ability to auto-index the oscillation data has considerably enhanced the usability of this method.

Theory

The basic theory in the rotation/oscillation method is that if a crystal is mounted such that it can be rotated about a direct crystal lattice (*a*-axis) and therefore perpendicular to the rotation axis (Figure 4.51). It is clear from the concept of reciprocal lattice that $b*c*$ reciprocal plane is perpendicular to the direct *a*-axis and hence perpendicular to the axis of rotation. This plane contains all the reflections from which the *h* index is zero, and the planes for which *h* = 1, 2, and so on are parallel to it and also perpendicular to the axis of rotation. As the crystal is turned (say 30°), these reciprocal planes are regarded as turning along with it and cutting the sphere of reflection (Figure 4.52).

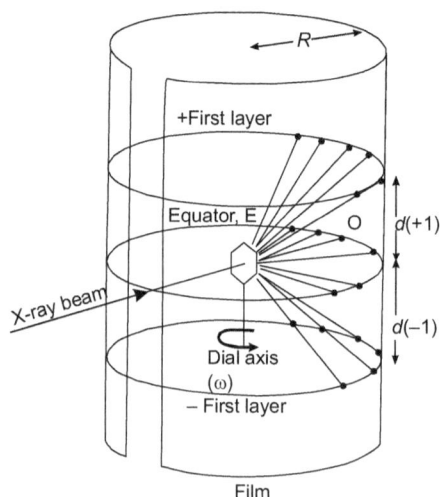

Figure 4.51 Basic geometry of the diffraction method, showing how diffraction spots are recorded on a cylindrical film placed around the crystal

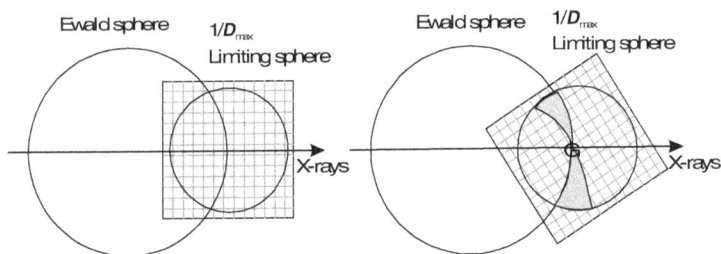

Figure 4.52 Reciprocal planes intersecting the Ewald sphere after rotating the crystal

After a 30° counterclockwise rotation of the crystal, starting from the left-hand side situation, all reflections in the shaded area will have passed the Ewald sphere and, thus, will have diffracted. As the crystal rotates, successive lattice points along a lattice line will be brought into reflecting position. The origin of the reciprocal lattice, point 000, is always in contact with the reflecting circle for the zero level, but as this also lies in line with the direct X-ray beam, its reflection cannot be measured directly.

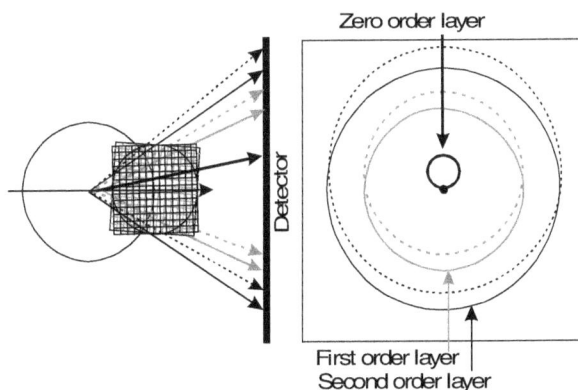

Figure 4.53 Zero order and upper level reflections

To measure the upper level reflections, the fact that any ray passing through a point on the circle cut by one of the upper level planes has the same vertical angle is important. Figure 4.53 shows the schematic depiction of an oscillation method diffraction pattern. Solid lines indicate the location of diffraction spots at the beginning of exposure, whereas the dashed lines give the situation at the end of the

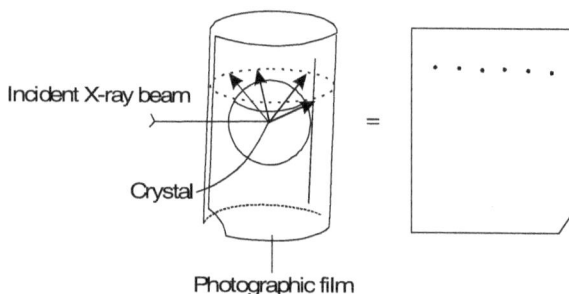

Figure 4.54 Recording a single layer line on cylindrical film

exposure (after a 5 degree counterclockwise rotation). The left-hand side shows the Ewald sphere construction and the right-hand side shows how the diffraction pattern will look like on the detector.

Thus, if the crystal is surrounded by a cylinder of film as shown in Figure 4.54, the extended rays will all intersect the film at the same height and the resulting spots will lie on a straight line when it is unrolled.

Oscillation Photographs

Rotation and oscillation photographs are generally used to align the crystals and to measure the cell edge along the axis of rotation. In some cases preliminary symmetry information can be obtained. An example for oscillation photograph is shown in Figure 4.55. The layer-line spacing, measured with respect to the zero layer, are denoted by $d(\pm n)$, where $d(+n) = d(-n)$, and are related to the repeat distance in the crystal parallel to the oscillation axis.

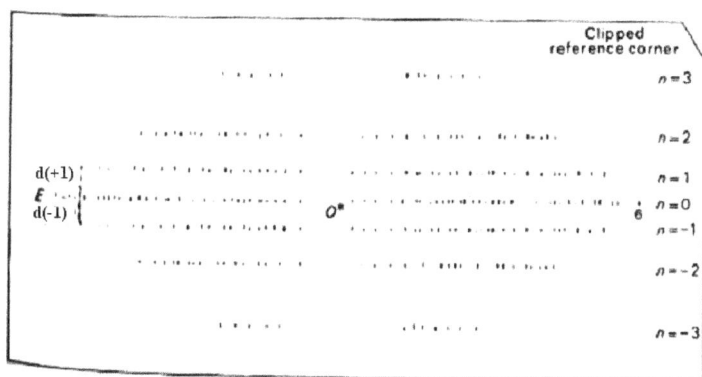

Figure 4.55 *a*-axis oscillation photograph of 15° oscillation of an orthorhombic crystal

Let the oscillation axis be a, and let R be the radius of the film, measured in the same units as d. The diffraction from the crystal gives rise to spectra of order ±1, ±2,..., $\pm n$; α_n is the scattering angle for the nth order maximum, measured with respect to the direct beam. The separation d_n between the equator and the nth layer line gives the periodicity along a, assuming the crystal to film distance R and the wavelength λ are known:

$$\frac{d_n}{R} = \tan\alpha_n$$

$$\alpha_n = \tan^{-1}\left(\frac{d_n}{R}\right)$$

$$\frac{n\lambda}{a_1} = \sin\alpha_n = \sin\left\{\tan^{-1}\left(\frac{d_n}{R}\right)\right\}$$

Here, $a_1 = \dfrac{n\lambda}{\sin\left\{\tan^{-1}\left(\dfrac{d_n}{R}\right)\right\}}$

For a known wavelength, this equation provides a convenient and reasonably accurate method for determining unit cell spacing. Oscillation photographs have several useful symmetry properties. A horizontal mirror line along the equator of a general oscillation photograph indicates a mirror plane perpendicular to the oscillation axis in the corresponding Laue group of the crystal. More observations on the symmetry of the Laue group can be made by arranging a particular crystal symmetry direction to be parallel to the X-ray beam at the midpoint of the oscillation range. To index the *a*-axis oscillation photograph, in zero level, the reflections are of the type *0kl*. The relevant portion of the reciprocal lattice is the Y^*Z^* net which, for an orthorhombic crystal, is determined by b^* $(= \lambda/b)$, c^* $(= \lambda/c)$, and a^* $(= 90°)$.

Limitations of the Method

While attempting to interpret the rotating-crystal photographs, such photographs are defective in two respects. Qualitatively, they yield no direct symmetry information about the crystal whatsoever, and quantitatively they are incapable of supplying sufficient information for indexing the reflections. Both of these deficiencies are so serious that they render the rotating crystal method a weak vehicle for crystal-structure analysis. Every rotating-crystal photograph has symmetry 2 mm when the angle between the rotation axes and the X-ray beam is 90°. This symmetry is not concerned with the crystal symmetry but rather with the fact that every stack of planes oblique to the rotation axis satisfies the Bragg's reflection condition four times on a single rotation of the crystal; this occurs when the common normal to the planes reaches the film in the first, second, third and fourth quadrants about the X-ray beam. The orientation error in mounting the crystal also leads to misinterpretation.

The indexing of the reflections on a rotating-crystal photograph is inherently impossible because there are three unknowns h, k and l to be determined, but the photographic film on which these three variables occur has only two dimensions. When the crystal is rotated about a crystallographic axis, then the index (corresponding to the axis about which the crystal is rotated) is fixed as a specific integer n because this is the common index for all reflections on the nth-order Laue cone. Thus, if the crystal is rotated about the c-axis, all reflections with a common index l occur along generators of the same Laue cone and so record on the same layer line. If the spot occurs on the nth layer line, its index is hkn. The other two indices h and k are indeterminate since these two variables are functions of the one measurable parameter along the layer line.

This problem of indexing rotating-crystal photographs was completely solved by Karl Weissenberg by spreading

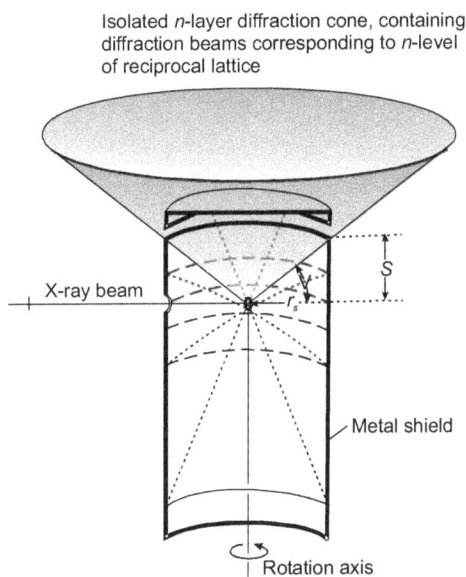

Isolated n-layer diffraction cone, containing diffraction beams corresponding to n-level of reciprocal lattice

Figure 4.56 Elimination of all Laue cones except one, by means of a layer-line screen

the two unknown variables *(h* and *k* when rotating a crystal about its *c*-axis) of the spots of one specific layer line over the two dimensions of the film. He arranged to screen out all but one layer line, as in Figure 4.56 by letting the diffracted rays, corresponding to this layer, reach the film through an annular slit in a cylinder placed co-axial with the film.

This amounts to recording only one Laue cone, say the *n*th. The directions of the reflections *hkn* occur along different generators of this cone, but generally each different reflection occurs as the crystal is in a different specific orientation ϕ about the rotation axis, such that the Bragg's reflection condition is satisfied for that d_{hkn} and later Weissenberg camera came into use.

PRECESSION METHOD

As the name suggests, precession photography involves making a crystal precess at a fixed angle around a defined axis. The precession camera was invented by Buerger in about 1940. This technique uses a flat film holder linked to the crystal oscillation axis (Figure 4.57). With the instrument set at zero, the X-ray beam must strike the crystal parallel to a real axis and perpendicular to the film. The crystal (and film) is tilted by an angle $\bar{\mu}$ of up to 30°, and allowed to 'precess' so that the real crystallographic axis traces a cone about the X-ray beam. The original version of the precession method was the consequence of an attempt to correct the lack of information about the symmetry of the crystal revealed by a rotating-crystal photograph. The symmetry in the photograph shows a projection of the rotating motion of the crystal.

The motion is complex but results in a photograph that gives an undistorted picture of the 'reciprocal lattice'. This is the second most common single-crystal camera in current use. It is very popular for studies of crystalline proteins.

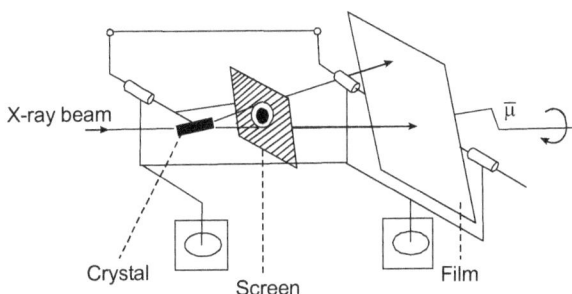

Figure 4.57 Buerger precession instrument

From these photographs the unit cell can be determined with ease and considerable accuracy. The symmetry and missing reflections of such photographs yield the diffraction symbol, which is the maximum qualitative information that can be determined from a set of X-ray photographs.

Buerger's Mark II Precession Instrument

The diagram of the three-dimensional linkage used to maintain the photographic film perpendicular to ON in the second version of the Mark II precession instrument is shown in Figure 4.58. The crystal and the film are suspended in gimbals so that the centre of the crystal and the centre of the zero-level position of the film remain unmoved and in line with the X-ray beam. The proper precession mechanism has two functions. It must make the crystal to undergo precessing motion, and it must cause the photographic film to execute a duplicate motion. The normal to the reciprocal lattice plane is the line ON. The plane normal ON takes a precessing motion because one end of it is carried around the drive shaft by an arc-shaped arm. The orientation of the film is directly controlled by the arc setting, the crystal then being made parallel. $O'N'$ is the arc radius. All features of both gimbals are kept parallel and the distance between them is fixed; the angular motions of the two suspensions about these two axes can be coupled by a single connecting link of length vv' attached by pins to the two suspensions. The motions of the horizontal axes H and H' also coupled regardless of the angular position of the suspensions about their vertical axes. The layer-line screen holder is permanently attached to the linkage which causes the precessing axis of the crystal and the film normal to be parallel. The layer-line screen which is usually a thin brass plate inserted in a holder is always normal to the precessing axis. The angular adjustments of the crystal and the dial adjustments permit orienting the selected translation of the crystal parallel to the axis of the nest of Laue cones defined by the centre of the layer-line screen holder. In order to take precession photographs of upper levels it is necessary to set the layer-line screen at the proper distance and to set the film forward by the proper distance.

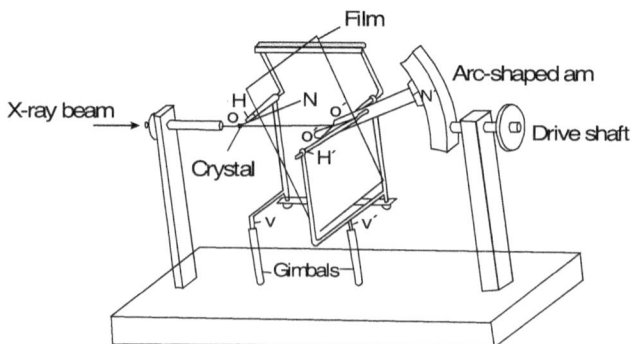

Figure 4.58 Diagram showing the coupling of the precession motions of the film and crystal

The basic idea behind this is, if a crystal is oriented so that its a-axis is parallel to the X-ray beam and a photograph is taken while the crystal is oscillated from this position by ω degrees, then the central portion of the film will show the $0kl$ reflections uncontaminated with other spots, provided ω

is sufficiently small. Buerger has shown that ω_{max}, the largest permissible value of ω, is given by

$$\sin^2 \omega_{max} = \lambda d^* \Big/ 2$$

where, λ is the wavelength in Å, and d^* is the interplanar spacing in the beam direction in $Å^{-1}$. Unfortunately, in most cases ω_{max} is so small that only a very limited area of the reciprocal net can be photographed. This area can be increased by a factor of about four by using a moving layer-line screen. The layer-line screen which is attached to the rotation axis has two apertures. The smaller semicircle on the left (Figure 4.59) eliminates all but the $0kl$ reflections which therefore appear on the left side of the film, while the larger semicircle admits only the $1kl$ reflections, which appear on the right of the photograph. Figure 4.59 shows that the 0-level screen begins to cut out the 0-level reflections when $\cos \omega = 1 - \lambda d^*$.

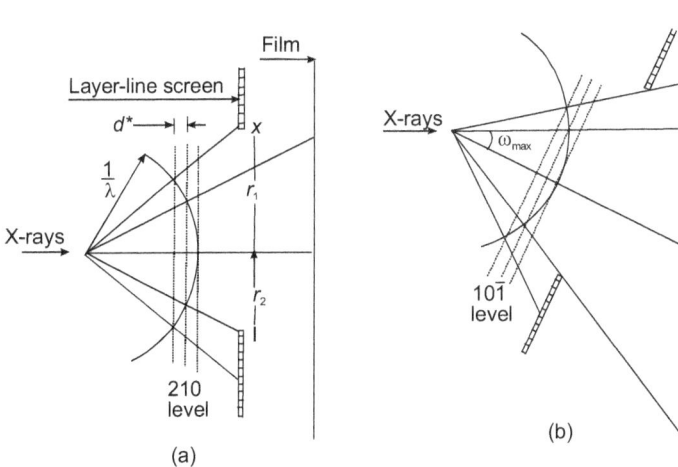

Figure 4.59 The sphere of reflection and the reciprocal lattice. (a) when $\omega = 0$ and (b) $\omega = \omega_{max}$

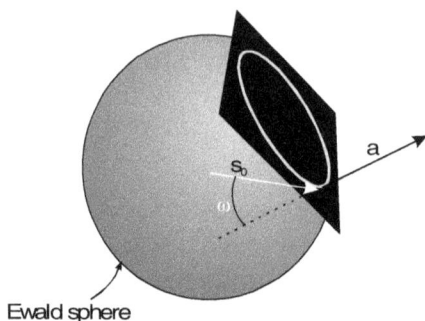

Figure 4.60   Precession camera geometry

Figure 4.60 shows if the angle of the beam with the a-axis is ω, then the correct angle setting for the screen is ω and by using a screen we can see one lattice plane is masked out.

Figure 4.61 shows a precession photographic method for one complete reciprocal plane ($\boldsymbol{b}^{*}\boldsymbol{c}^{*}$ plane at $b = 0$) on a film.

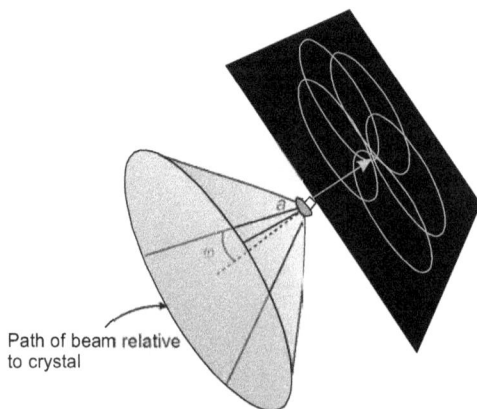

Figure 4.61   Precession photography

If the crystal is precisely aligned such that a real space unit-cell axis lies along the rotation axis, a precession photograph can be arranged to provide view of a single plane through diffraction space. Since this involves introducing a metal-layer screen that blocks most of the diffraction and only allows passage of that from the desired layer. The method produces an undistorted view of a single reciprocal lattice plane. In the (very) old days we used to compare zero-level projections (*0kl, h0l, hk0*) between natives and potential heavy atom derivatives to look for relative intensity changes. These days we can do the same thing in a fraction of the time using conventional oscillation photography.

Applications of Precession Photographs

Precession photographs are used in the crystal structure analysis for two main purposes.

1. to permit symmetry observations leading to determination of the space group measurements leading to knowledge of the determination of unit-cell.

2. to provide data for the measurement of the intensities of the $|F_{hkl}|$s.

Interpretation

A typical precession photograph is shown in the Figure 4.62. Note the systematic absences, relative spaces of spots and angle between the axes.

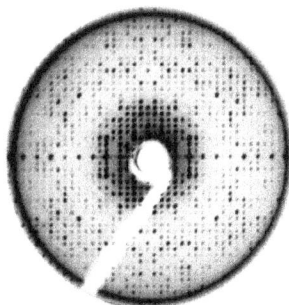

Figure 4.62   Zero-level precession photograph

Diffraction symbol   All the qualitative information which can be derived from a set of X-ray diffraction photographs can be concentrated into a short sequence of symbols which together constitute the **diffraction symbol**. A diffraction symbol is determined by the Friedel symmetry, the lattice type, and the record of the direction of glide planes and screw axes. There are 122 such symbols, and they are of great utility in the determination of the space group of a crystal. The determination of the diffraction symbol is especially easy from precession photographs.

Freidel symmetry   Friedel symmetry can be determined by inspection from an appropriate set of precession photographs. Each rational direction of the crystal has one of the possible symmetries about a line in three-dimensional space.

Screw axis and glide plane   The easiest way to find the screw axis and glide planes is to compare the zero level with one upper level of the reciprocal lattice parallel to the symmetry element in question. The possible pattern of missing points (systematic absences) reveals the corresponding space group. Some of the examples are shown below.

orthorhombic + odd h reflections missing in $h00$ line → 2_1

trigonal + only $l = 3n$ reflections present in $00l$ line → 3_1 or 3_2

hexagonal + only $l = 2n$ reflections present in $00l$ line → 6_3

tetragonal + only $l = 4n$ reflections present in $00l$ line → 4_1 or 4_3

Enantiomeric space groups ($P4_1$ and $P4_3$) cannot be distinguished until the phases have been solved.

Unit cell determination   Unit cell type is determined by noting the edge lengths and interaxial angles of the reciprocal cell from the precession photograph. The results can be readily transformed into edge lengths and interaxial angles of the direct cell by using standard relations between the two cells. To express the linear elements of the direct cell in Å, a proportionality constant should be added with those of reciprocal cells in Å^{-1} units. Generally, in diffraction experiments a constant λ is used and the precession instrument introduces a magnification factor M, so that measurements made on precession photographs involve a constant λM.

Intensity measurement   Precession method is more convenient to obtain an intensity data for a three-dimensional study, because the entire reciprocal lattice within recording range can be surveyed with a single mounting of the crystal. To obtain all the reflections in a single mount, a translation of the reciprocal lattice should be set parallel to the dial axis. The indexing procedure is much simplified if this translation is an edge of the reciprocal cell. If the crystal has a unique axis of high Friedel symmetry, this should be set parallel to the dial axis.

As the precession method provides a simple and direct way of establishing the space group and unit cell of a crystal, the method came into use in many laboratories. This type of instrument was used in the early days of protein crystallography before advanced algorithms for auto-indexing oscillation photographs which made the interpretation of those more straightforward.

X-RAY DIFFRACTOMETER

Nowadays the Bragg intensities from the diffraction of the crystal are collected using single-crystal X-ray diffractometer. The basic components of a typical single-crystal X-ray diffractometer are:

1. X-ray source
2. Goniometer system
3. Video camera or microscope

4. X-ray detector system
5. Microprocessor-based interface module
6. Host computer

X-ray Source

To generate X-rays, the electrons are accelerated by an electric field (about 30 kV) against a metal target, which slows them rapidly by collisions. Copper or molybdenum are used as targets. Copper K_α is the conventional choice to use with organic crystals as it penetrates sufficiently and also does not suffer too badly from absorption in the crystal or while passing through the air. As scintillation counters have nearly perfect counting efficiency for Mo K_α, it is also used in different diffractometers.

Goniometer

Goniometer system allows the specimen to be precisely oriented in any position while remaining in the X-ray beam. The most critical mechanical component in an X-ray diffractometer system is the goniometer assembly, which must be capable of keeping the specimen centered in the incident X-ray beam while at the same time changing its orientation in order to collect many thousands of frames of data in reciprocal space. There are 2-circle, 3-circle and 4-circle goniostats used in diffractometer systems. Figure 4.63 shows the 3-circle and 4-circle goniometer. The two-circle is most limited, consisting of a rotation axis and a swing movement for the detector. A three-circle goniometer has a rotation axis and a ϕ rotation mounted with χ fixed at usually 45°. A swing angle is also provided for the detector. Data are usually collected by rotating around the rotation axis as far as possible. The crystal can be rotated around the ϕ axis to collect new data. Usually, rotating ϕ 90° will give the most new unique data. A four-circle goniostat will allow the most control over data collection. It has all the movements of the three-circle, and χ can be adjusted a full 360°.

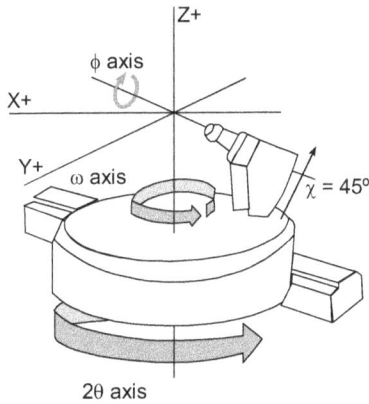

Three-circle goniometer

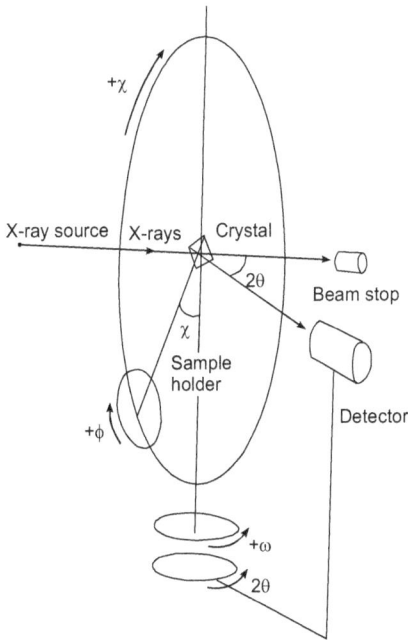

Four-circle goniometer

Figure 4.63 Goniometer

A four-circle diffractometer consists of a four-circle goniostat. The four circles are ω, χ, ϕ and 2θ. By moving the angles, it is possible to bring every reflection into diffracting conditions such that the reflections can be counted. Three-circle goniostat has a fixed χ angle usually (45°) and two circles $\chi = 0$ and $\phi = 0$.

Crystal mounting A metal pin which is suitable for insertion into the goniometer head is used for crystal mounting which is shown in Figure 4.64. A very fine glass fibre (20–70 μm in diameter) is pasted with sealing wax to the metal pin (usually brass) before mounting the crystal.

Figure 4.64 Goniometer head

The crystal is attached to the fibre tip by an adhesive. Crystals that are unstable in air are mounted in capillaries (Quartz or Lindemann glass) sealed with sealing wax and mounted in turn in brass pin. Once it is glued, the pin is securely attached to a goniometer head (sample holder) in an arbitrary orientation. The goniometer head is then placed at the base of the goniometer assembly and the crystal is optically aligned in the centre of the incident X-ray beam using a video camera or microscope. The orthogonal X, Y, and Z translations on the goniometer head are adjusted until the specimen is centered on the cross hairs for all crystal orientations.

Video Camera or Microscope

Video camera or microscope is used for aligning the specimen and indexing the crystal faces. The crystal is optically aligned in the centre of the incident X-ray beam using a video camera or microscope.

X-ray Detector System

The most important and most expensive component of any modern single-crystal diffractometer system is its detector system. The detectors are of various types such as scintillation counters, area detector, charge-coupled device, image plate, etc. Before 1994, the diffractometers were with scintillation type of detecting systems. **CAD-4 diffractometers** (computer assisted diffractometer-circle 4) consist of scintillation counters.

Area detectors are the most commonly used detectors in data collection technique. Area detector collects a larger volume of reciprocal lattice than other type of detectors and so it is more efficient. An area detector can be equipped with three types of goniometers.

Most new instruments purchased since 1994 use a detector system based upon **CCD** (charge-coupled device) technology (Figure 4.65). These detectors include a square CCD-microprocessor chip with physical dimensions ranging from 24 mm (1K-chip) to 61 mm (4K-chip) on each side. Each of these CCD chips contains from 1024×1024 (1K-chip) to 4096×4096 (4K-chip) independent pixels. The CCD chip is usually bonded to a fibre-optics taper to increase the effective size to 61 mm on a side (Mo target chemical crystallography systems) or 90 mm on a side (Cu target protein crystallography systems). The front end of the fibre-optics taper is attached to a phosphor screen, optimized for either Mo or Cu radiation, to convert the X-ray photons to visible wavelength photons which can be transmitted to the CCD chip by the fibre-optics bundle. In order to reduce background noise and improve counting statistics, the CCD chip must be cooled to about −45°C using Peltier cooling methods. Typical CCD detector systems (Figure 4.65) have counting efficiencies ranging from 10 to 90 electrons per X-ray photon.

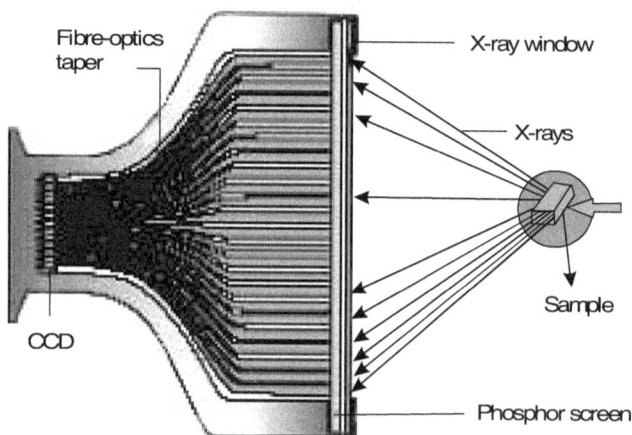

Figure 4.65   CCD (charge-coupled device) based detector

Image plates   Image plates can be used as a replacement of the area detector. Image plates (Figure 4.66) are exposed with X-rays, as with any other detector, and the X-ray photon causes a chemical change in the plate coating that releases a fluorescence which is detected by a photomultiplier when scanned with light of proper wavelength. Image plates are read out by a laser beam on a scanner. Image plates have a wider range of sensitivity with respect to X-ray wavelengths, which gives them higher counting efficiency at higher energies.

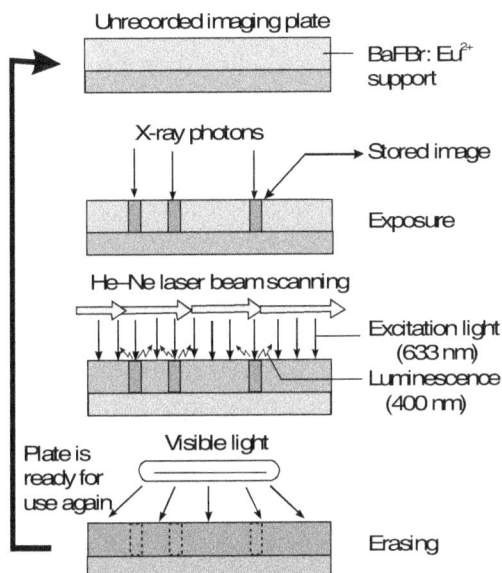

Figure 4.66 Image plate

Microprocessor-based interface module A microprocessor-based interface module receives commands from a host computer and carries out all real-time instrument control functions to drive goniometer motors, monitor the detector system, open and close the shutter and monitor collision sensors, and safety interlocks.

Host Computer

Modern diffractometers come with software packages that enable them to carry out automatic searches of a portion of reciprocal space, identify a number of reflections, center these accurately in the detector aperture and measure the setting angles, and then calculate probable cell parameters as well as the orientation matrix required to collect a data set. A host computer with a large hard disc mass storage device, a video monitor and keyboard, and diffractometer is to control programs, to control the data collection strategy and to send commands to the microprocessor. Structure determination calculations may be carried out on the computer used for data collection or they may be performed on a second computer linked to the diffractometer system.

DATA COLLECTION

A preliminary rotational image is then collected for one minute with the CCD detector to screen the specimen for analysis and to select suitable parameter values for subsequent steps. In order to determine the unit cell, a preliminary set of frames is measured using an automatic routine. For example, three sets of frames are collected in different parts of reciprocal space. These frames are then processed to locate spots on individual frames and to then determine the centers of reflections. An auto-indexing routine selects the appropriate reduced primitive unit cell and calculates the corresponding orientation matrix and lattice constants. This preliminary unit cell is then refined using a non-linear least-squares algorithm and converted automatically to the appropriate crystal system and Bravais lattice. This new cell is again refined by the non-linear least-squares algorithm to yield an accurate orientation matrix which may be used to index crystal faces and to carry out integration calculations after intensity data collection.

After the above geometric data collection steps have been completed and an accurate orientation matrix has been calculated, intensity data collection is carried out. Typically, a sphere or hemisphere of data is collected using a narrow-frame scan method in which several sets of frames (runs) are collected by scanning in 0.1° to 0.3° increments in the ω and/or ϕ angle, while keeping all other instrument angles constant. There are options to limit data to a unique set of reflections, thereby reducing data collection times for high-symmetry crystal systems. Each two-dimensional frame is a two-dimensional array of independent pixels, each of which has an almost infinite dynamic range for sensitivity. Even though, the microprocessor chip in the CCD camera contains from 1024 × 1024 (1K) to 4096 × 4096 (4K) independent pixels, the software usually reads the data out in 'binned' mode as 512 × 512 or 1024 × 1024 frames.

A complete data set may require a few hours to overnight depending upon the size of the specimen and its diffracting power. Typical exposure times are 10 to 30 seconds per frame. About 4 to 8 hours are needed for total data collection. When the complete set of frames has been collected for a given specimen, the entire data set must be processed to obtain accurate integrated intensities for individual reflections. This process includes corrections for instrumental factors, polarization effects, X-ray absorption and possibly crystal decomposition. The integration process reduced the raw frame data, which require from 500 to 2000 megabytes of disc storage to a small set of individual integrated intensities for individual reflections. The final unit-cell constants are calculated from the centroids of many thousands of reflections selected from the entire data set and typically have relative errors of less than 1/10,000. Once the structure amplitudes are known, the phase problem must be solved to find a self-consistent set of phases that can be combined with the structure factor amplitudes to obtain the electron density and thereby determine the structure of the crystal.

Determination of Structure Factor Amplitudes from Intensities

Data collection yields a set of peak counts at discrete measurement points identified by three indices denoted h, k and l (which are also known as Miller indices). To obtain the structure factor amplitudes, the first step is to reduce the intensities. This process includes adjustments for mechanical design of the diffractometer as well as various corrections to the data.

The intensity of an X-ray beam is proportional to the square of the wave amplitude. But the measured intensity is affected by various factors, however, for which corrections must be applied. The conversion of intensities I to "observed structure factor amplitudes" $|F_o|$ and, correspondingly, of s.u.'s $\sigma(I)$ to $\sigma(F_o)$ is known as data reduction. The factors which affect the intensities are mainly partial polarization, extinction, absorption, thermal motion, etc.

There are corrections associated with data collection process, which are geometrical in nature. These are a function of the geometry of the equipment and so are instrument dependent. The reflected radiation is partially polarized and for this Lorentz-polarization correction is applied. A correction may also be needed for changes in the incident X-ray beam intensity or in the scattering power of the crystal during the experiment. The incident X-ray beam correction is important for synchrotron radiation, which decays gradually and significantly and can be directly monitored. The changes in the scattering power may be caused by decomposition of the sample in the high-energy X-ray beam.

Data Reduction

In some cases, the diffractometer software performs initial calculations that produce raw intensities. In other cases like Enraf–Nonius CAD-4, the essential information is provided but the calculation must be done separately. For CAD-4, let

C be the total count

R be the ratio of scan time to background counting time (usually 2 for CAD4)

B be the total background count

N be the ratio of fastest possible scan rate to scan rate of the measurement

A be the attenuation factor

S be the actual scan speed

Then the intensity is given by

$$I_{hkl} = \frac{S \cdot A}{N} \left(C - (R \cdot B) \right)$$

These are all the adjustments during data collection.

Extinction correction In the data reduction process, the correction for extinction (i.e., a phenomenon which results in the attenuation of the primary beam of X-radiation when the crystal is in diffracting position and so reduces the intensity of the diffracted beam) is to be carried out. This effect is well known theoretically, but is very difficult experimentally because it depends on the physical perfection of the crystal. For this reason, it is customarily ignored in many current crystal structure analysis procedures.

Lorentz-polarization correction Lp is the combination of two geometric correction factors. Lorentz factor corrects for the difference in time required for a reciprocal lattice point to pass through the sphere of reflection. It depends on the method of data collection and for the usual 2θ and ω scan methods used with diffractometers, the Lorentz factor is

$$L = \frac{1}{\sin 2\theta}$$

Because the X-ray beam is partially polarized, its reflection efficiency varies with the reflection angle. The polarization factor is independent of the method of data collection and when no monochromator is used, it is simply

$$p = \frac{1 + \cos^2 2\theta}{2}$$

Absorption correction Crystals absorb X-rays thereby causing a reduction in the intensities of the reflection data. Each reflection is affected differently by absorption, because the absorption depends on the path length of the X-rays through the crystal, and this varies as the crystal orientation is changed. The amount of absorption is a function of the path length through the crystal and the absorption coefficients of the atoms making up the crystal.

$$I = I_o e^{-\mu t}$$

where,

   μ is the absorption coefficient,

   t is the thickness of the crystal, and

   I_o is the incident intensity.

Absorption can affect the accuracy of the bond distances and angles and cause unusual shaped thermal ellipsoids of the atoms, inflate the agreement factors, and in some cases inhibit the ability to solve the structure.

Various types of absorption correction procedures available are

Psi scan Empirical method based on measurements of absorption as a function of the orientation angle ψ.

Gaussian Numerical method based on measurements of the crystal, identification of the crystal faces, mathematically dividing the crystal into a three-dimensional grid and summing the path length through the crystal.

Analytical Analytical method based on same measurements as Gaussian method but, the calculation is made by summing the path length through a collection of polyhedrons conforming to the crystal shape.

Spherical   Numerical method based on measurement of the radius of the crystal, assumed to be nearly spherical in shape.

Fourier   Empirical method based on minimizing the residual difference between observed and calculated intensities.

The various corrections for the intensities are applied also to their s.u.'s. The result of this whole process, which usually takes only a matter of minutes on a computer, is a list of reflections as h, k, l, $|F_o|$ and $\sigma(|F_o|)$ or h, k, l, I and $\sigma(I)$.

The data reduction process also includes the merging and averaging of repeated and symmetry-equivalent measurements in order to produce a unique, corrected and scaled set of data. This calculation affords a numerical measure of the agreement among equivalent reflections, which is one indication of the quality of the data and the appropriateness of the applied corrections.

STEPS IN CRYSTAL STRUCTURE DETERMINATION

The various steps involved in the three-dimensional crystal structure determination are

1. Crystallization

2. Crystal mounting

3. Collection of Bragg intensities

4. Space group determination

5. Data reduction

6. Structure solution

7. Least squares refinement

8. Obtaining correct structure

9. Structure analysis

Crystallization

The compounds synthesized in chemical laboratory or obtained from any natural products can be crystallized after isolation and purification by any of the chromatographic techniques. There are several methods to crystallize the small molecules and the simplest among those is slow evaporation technique. Generally, crystallization requires supersaturation of the sample in the solvent, so that the system gradually proceeds towards an equilibrium state in which the compound is partitioned between a solid and a soluble phase. If the various types of intermolecular interactions reduce the free energy of the system, then the molecules will tend to crystallize. Supersaturated solutions that result in the removal of some of water may give rise to situations where the molecules may aggregate as an amorphous precipitate or form crystals. Formation of amorphous precipitate implies that saturation is too extensive or has proceeded too rapidly. It is therefore essential to approach the point of inadequate solvation very slowly to get crystals.

The supersaturated solution (sample + solvent) is taken in a beaker and it is covered by a thin parafilm, in which a small hole is made that enables slow evaporation. The set-up is left undisturbed. Depending upon the nature of the sample, some may take few hours or days to crystallize. The obtained crystal should be transparent enough and should be having the minimum dimension as 0.2 mm × 0.2 mm × 0.2 mm. For collecting good Bragg's reflections by X-ray diffraction experiment, the crystals free from defects or twinning are required. Hence the purity of the sample is also very important so as to obtain good quality crystals.

Crystal Mounting

After optical examination, stable crystals of suitable size are mounted, using adhesive, on to a glass fibre, while unstable crystals, such as proteins, are mounted in the presence of mother liquor inside the sealed capillary tubes. For precise data collection, a crystallographic axis should be perpendicular or closely perpendicular to the fibre or glass capillary. The final alignment of the axis can be undertaken with the crystal mounted on the camera using the goniometer arcs. The camera is fitted with a telescope by means of which the crystal can be centred, that is, adjusted on the axes such that it rotates within its own volume, and the desired axis aligned approximately along the X-ray beam.

Collection of Bragg's Intensities

X-ray diffraction data can be collected using three different diffractometers, namely, Bruker Apex CCD area detector, Enraf–Nonius CAD-4 and SMART CCD area detector. A crystal of suitable size is mounted and accurate values of the unit cell parameters are obtained by least-squares analysis of the values for several high Bragg angle reflections measured on a diffractometer equipped with MoK_{α} or CuK_{α} radiation. Intensity data are collected at room temperature. For diffractometer data, all the intensities were corrected for variable scan speed, background and attenuation using the relation,

$$I_{raw} = f[N_c - 2(Lb + Rb)]NPI \qquad (4.62)$$

where,

I_{raw} is the relative intensity,

N_c is the peak count,

Lb and Rb are the left and right background counts respectively,

NPI is the scan speed parameter and

f is the attenuation factor.

Data Reduction

The observed structure factor for each reflection is obtained using the equation,

$$F_O = K \left[\frac{I_{raw}}{Lp} \right]^{1/2} \qquad (4.63)$$

where,

K is the scaling factor relating the arbitrary intensity counts to the absolute value of the structure amplitude,

L is the Lorentz factor = $1/\sin 2\theta$, where, θ is the Bragg's angle of reflection and

p is the polarization factor = $\dfrac{1 + \cos^2 2\theta}{2}$.

In X-ray diffraction spectra, it is assumed that the crystal structure is a static one, i.e., it can be thought as a periodic pattern of stationary atoms. Actually all the atoms are in thermal motion at different temperatures and the electrons of each atom sweep out a larger average volume than they would occupy if the atom were at rest. This causes the effective *f* curves of the atoms to fall off more rapidly with $(\sin \theta)/\lambda$ than for the same atoms at rest. As each atom undergoes a motion, its electron density is smeared over a small anisotropic volume, usually regarded as a triaxial ellipsoid in the general case. Thus a Debye–Waller temperature correction is applied to the scattering factor, which is given by

$$f = f_0 e^{-(B \sin^2 \theta)/\lambda^2} \qquad (4.64)$$

The quantity *B* is called the thermal factor and its value is

$$B = 8\pi^2 \overline{U^2} \qquad (4.65)$$

where $\overline{U^2}$ is the mean square displacement normal to the reflecting plane of the atoms from their mean position. This correction is used for each f in the summation for structure factor.

Determination of Thermal and Scale Factors

The temperature parameter, B, can be found, and at the same time the set of F's can be placed upon an absolute basis, by a method first presented by Prof. A.J.C. Wilson (the father of crystallographic statistics) and later rediscovered by Prof. David Harker. The theory is as follows:

$$F_{hkl} = \sum_{j=1}^{N} f_j e^{2\pi i(hx_j + ky_j + lz_j)} \tag{4.66}$$

The square of the absolute value of F is found by multiplying it by its complex conjugate.

$$\begin{aligned}
\left|F\right|^2 &= FF^* \\
&= \left(\sum_i f_i e^{2\pi i(hx_i + ky_i + lz_i)}\right)\left(\sum_j f_j e^{-2\pi i(hx_j + ky_j + lz_j)}\right) \\
&= \sum_i \sum_j f_i f_j e^{2\pi i(h[x_i - x_j] + k[y_i - y_j] + l[z_i - z_j])}
\end{aligned} \tag{4.67}$$

This can be separated into two parts, according as $i = j$ and $i \neq j$:

$$|F|^2 = \sum_j f_j^2 + \sum_i \sum_j f_i f_j \; e^{2\pi i(h[x_i - x_j] + k[y_i - y_j] + l[z_i - z_j])} \tag{4.68}$$

If either the average or the sum is taken over all hkl, the second term tends to zero since it contains as many positives as negative components, so that

$$\left|\overline{F}\right|^2 = \sum_j \overline{f_j^2} \tag{4.69}$$

Now, consider $\left|\overline{F_{\text{obs}}}\right|^2$, the absolute value of $\left|\overline{F}\right|^2 \cdot \left|\overline{F_{\text{obs}}}\right|$ is usually known on arbitrary scale only, so

$$\left|F_{\text{obs}}\right|^2 = K\left|F\right|^2 \tag{4.70}$$

Accordingly,
$$\left|\overline{F_{\text{obs}}}\right|^2 = K\left|\overline{F}\right|^2 \tag{4.71}$$

But the right side of equation 4.71 is known from equation 4.69. Thus

$$\left|\overline{F_{\text{obs}}}\right|^2 = K\sum_j \overline{f_j^2}$$

From this the scale factor, $K = \dfrac{\left|\overline{F_{\text{obs}}}\right|^2}{\sum_j \overline{f_j^2}}$ $\tag{4.72}$

But the f_js are the true scattering powers under the conditions of observation. Assuming the Debye–Waller temperature correction,

$$f^2 = f_0^2 e^{-(2B\sin^2\theta)/\lambda^2} \tag{4.73}$$

Accordingly, equation 4.72 may be written as

$$K = \dfrac{\left|\overline{F_{\text{obs}}}\right|^2}{\sum_j \overline{f_{j0}^2}\, e^{-(2B\sin^2\theta)/\lambda^2}} \tag{4.74}$$

From this it follows that

$$\dfrac{\left|\overline{F_{\text{obs}}}\right|^2}{\sum_j \overline{f_{j0}^2}} = K e^{-(2B\sin^2\theta)/\lambda^2} \tag{4.75}$$

If this is plotted against $\sin^2\theta$ or $\sin^2\theta/\lambda^2$, then, as θ tends to 0,

$$\frac{\overline{\left|F_{\text{obs}}\right|^2}}{\sum_j \overline{f_{j0}}^2} = K \tag{4.76}$$

This provides a way of finding coefficient K, which is necessary to place the $\left|F_{obs}\right|^2$'s on absolute scale. If logarithms are taken on both sides of equation 4.75,

$$\ln\left(\frac{\overline{\left|F_{\text{obs}}\right|^2}}{\sum_j \overline{f_{j0}}^2}\right) = \ln K - \frac{2B\sin^2\theta}{\lambda^2} \tag{4.77}$$

and is plotted in Figure 4.67 which is known as **Wilson's plot**.

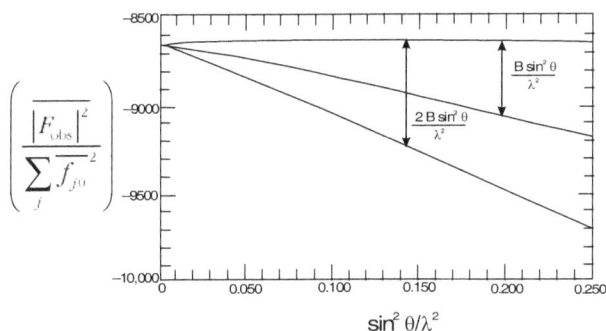

Figure 4.67 Wilson's plot

To solve equation 4.77 for B, it is convenient to rearrange it to

$$\ln\left(\frac{\overline{\left|F_{\text{obs}}\right|^2}}{\sum_j \overline{f_{j0}}^2}\right) - \ln K = -\frac{2B\sin^2\theta}{\lambda^2} \tag{4.78}$$

This has the form $y = Ax$ and to solve for A,

$$A = \frac{y}{x} = \frac{y_2 - y_1}{x_2 - x_1}$$

Similarly, $\dfrac{-2B}{\lambda^2} = \dfrac{\ln\left(\dfrac{\overline{|F_{obs}|^2}}{\sum_j f_{j0}^2}\right) - \ln K}{\sin^2 \theta}$ (4.79)

The intercept on the y-axis (c) gives ln K. Therefore, $K = e^c$; the slope of the straight line gives 2B.

Space Group Determination

The determination of space group of a crystal is an important feature in the crystal structure analysis. From the Bragg's intensities *hkl* collected from X-ray diffraction experiment, few important reflections are chosen and limiting conditions are deduced. Table 4.3 shows some reflection data for monoclinic crystal.

Table 4.3   Reflection data for monoclinic crystal

	100	204	111	322
	200	402	122	020
hkl	300	502	113	040
	400	110	311	060
	202	310	123	080

The reflection types are *h*00, *hkl*, *h0l*, *hk*0 and 0*k*0. There is no condition on *hkl*, *h*00, but we can deduce the following limiting conditions.

$$h0l \ (l = 2n)$$

$$0k0 \ (k = 2n)$$

These limiting conditions correspond to P2$_1$/c. In this way, by the systematic absences among all the collected Bragg's intensities, the space group of the particular crystal system can be determined. The limiting conditions for all the space groups of seven crystal systems are given in volume A of the *International Tables for Crystallography*.

Structure Solution

The structure solution can be obtained by any of the methods that determine the correct phases without any ambiguities. Here we deal with the structure solution by direct methods in which the phases are **directly** obtained from the observed intensities.

Example of crystal structure analysis of a small molecule The small molecules having some biological properties are synthesized in chemical laboratory and are crystallized by any of the crystallization methods such as slow evaporation, solution growth, etc. The Bragg's intensities are collected by X-ray diffractometer and the scaling and reduction of Bragg's intensities are done. (The crystal structure analysis of a synthesized small molecule using direct-method procedures is explained here.) The crystal structure is solved by using SHELXS 97 program and refined using least square refinement procedures SHELXL 97. Both the programs were written by Prof. George Sheldrick from Germany.

The chemical diagram of an example compound is shown in Figure 4.68. The total number of non-hydrogen atoms in the molecule is 18 only.

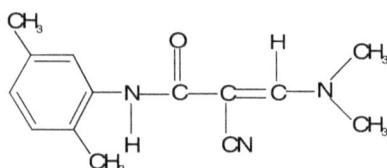

Figure 4.68 Chemical diagram

The input file for the SHELXS program includes instruction file (X.ins) and reflection file (X.hkl). The instruction file is created by the crystallographer, which includes the unit cell parameters, space group, scattering elements, etc. A typical instruction file is shown below.

```
TITL  P2(1)/n
CELL  0.71073 9.9315  11.8940  11.3979  90.000  97.079  90.000
ZERR  4    0.0002    0.0004    0.0004    0.000   0.014   0.000
LATT  1
SYMM  0.5-X,  0.5+Y,  0.5-Z
SFAC  C  H  N  O
UNIT  56  68  12  4
TREF
HKLF  4
END
```

The first line (TITL, P2(1)/n) indicates space group, CELL indicates the wavelength used and unit cell parameters (a, b, c and α, β, γ), the third line ZERR indicates the number of molecules in the unit cell (Z) and the errors in experimental measurement of parameters a, b, c and α, β, γ. LATT 1 indicates centrosymmetric (-1 indicates non-centro symmetric) and SFAC indicates the scattering species C, H, N and O. The UNIT represents the total number of atoms of corresponding species in a unit cell. There should be a one-to-one match between the SFAC and UNIT entries (order should not be changed!!). SYMM denotes the symmetry equivalent position corresponding to the particular space group. Original position (x, y, z) is assumed and the centrosymmetric equivalents are also assumed by the program. TREF is for tangent refinement method of deriving phases by direct method and HKLF 4 means the reflection file contains h, k, l, I and $\sigma(I)$. (HKLF 3 means h, k, l, $|F|$ and $\sigma(|F|)$. The reflection file (X.hkl) is as shown below.

h	k	l	I	$\sigma(I)$
0	1	−1	2319.05	89.78
0	1	1	2396.30	92.66
0	1	−2	1041.11	41.16
0	1	2	1028.53	40.47
0	1	−3	1763.57	68.69
0	1	3	1819.99	70.57
0	1	−4	563.07	22.95
0	1	−5	267.10	11.72
0	1	−6	280.31	12.19
0	1	−7	20.06	2.55
0	2	0	2.97	2.02
0	2	−1	709.65	28.82
0	2	1	692.93	28.11
0	2	−2	4959.22	190.87
0	2	2	5004.24	192.42
0	2	−3	10.66	3.03
0	2	3	9.93	2.47
⋮	⋮	⋮	⋮	⋮
⋮	⋮	⋮	⋮	⋮
⋮	⋮	⋮	⋮	⋮
0	0	0	0.00	0.00

With X.ins file along with X.hkl file, the solution is obtained by using SHELXS 97 program.

The sample output of the structure solution (X.lst) denotes that among 3291 unique reflections, 2073 are observed and the R_{int} value is found to be 0.0374 which indicates the good quality of data. The results of the structure solution is given by the program as X.res file and the program details are given in X.lst file. In direct methods, the strongest reflections are chosen (origin defining reflections) for generation of the phase relationships. Then various possible combinations of phases are tried, either by assigning values to a few reflections and using the probability relationships to improve them so that they fit the relationships better. The combination of phases giving the best agreement with the expected relationships is then used, together with the observed amplitudes, in a reverse Fourier transformation to calculate an electron density map, and this is examined for recognizable molecular features. In typical cases a few tens of initial phase sets are tried and several of these are likely to lead to a recognizable correct structure showing most or all of the non-hydrogen atoms. The X.lst indicates that SHELXS program has located a total of 18 atoms. Sample output of SHELXS program is shown below.

```
++++++++++++++++++++++++++++++++++++++++++++++++++++++++++++++++
+  SHELXS-97  -  CRYSTAL STRUCTURE SOLUTION  - WinGX VERSION  +
+  Copyright(C) George M. Sheldrick 1986-97      Release 97-2  +
+  s                         started at 16:42:09 on 05 Sep 2007  +
++++++++++++++++++++++++++++++++++++++++++++++++++++++++++++++++

Read instructions and process reflection data
Data:    3291 unique,   2673 observed     R(int) = 0.0374  R(sigma) = 0.0475
Systematic absence violations:     0    Bad equivalents:     1

ESEL  Emin  1.200    Emax  5.000    Del0 0.005    renorm 0.700    axis 0
OMIT  s  4.00    2theta(lim)  180.0
INIT  nn   11     nf   16    s+  0.300    s-  0.200    wr  0.200
PHAN  steps   10   cool 0.900   Boltz 0.400   ns  152   mtpr   40   mnqr  10
TREF  np       256.    nE   244    kapscal  0.900    ntan   2    wn -0.950
FMAP  code  8
PLAN  npeaks   -28    del1 0.500    del2 1.500
MORE  verbosity  1
TIME  t    9999990.

   152 Reflections and   1281. unique TPR for phase annealing
   244 Phases refined using    3980. unique TPR
   383 Reflections and    6872. unique TPR for R(alpha)
   823 Unique negative quartets found,   823 used for phase refinement
    92 Unique NQR employed in phase annealing
   128 Parallel refinements, highest memory -    4989 /   70800

  Try    Ralpha Nqual Sigma-1 M(abs) CFOM   Semivariants
1125009. 0.099 -1.757  0.544  1.070  1.137* ---+-- +-+-+ ---++ -+-++ ---+-
Freq: 0 26 0 0 0 0 0 0 0 0 0 0 28 0 0 0 1 4 0 0 0 0 0 0 0 0 0 1 4 0 / 128

 1677997. 0.009 -0.952  0.544  1.003  0.000  --++- +-+-+ ----++ -+-++ ---+-
Freq: 0 68 0 0 0 0 0 0 0 0 0 0 52 0 0 0 1 5 0 0 0 0 0 0 0 0 0 1 4 0 / 256

    256. Phase sets refined - best is code  1125009.  with CFOM -  0.0049

Fourier and peaksearch
RE = 0.287 for   18 atoms and   825 E-values
Fourier and peaksearch
RE = 0.284 for   18 atoms and   825 E-values
Fourier and peaksearch

++++++++++++++++++++++++++++++++++++++++++++++++++++++++++++++++++++++++
+  0                     finished at 16:42:18   Total CPU time:      0.0 secs  +
++++++++++++++++++++++++++++++++++++++++++++++++++++++++++++++++++++++++
```

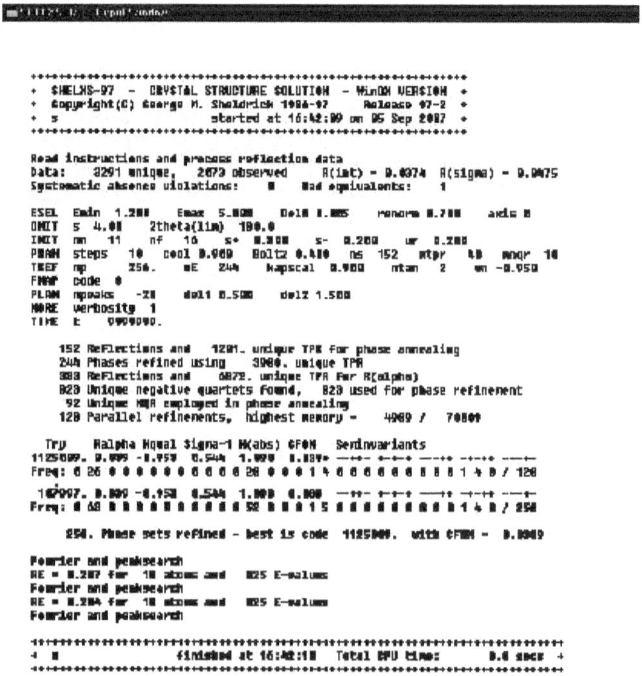

Sample output of SHELXS program

The E-map which is in the X.lst file is shown in Figure 4.69. With bond geometry and E-map and with the chemical knowledge, connectivity diagram can be deduced.

The initial structure solution has revealed positions for all the atoms (except for hydrogen atoms, which have very little electron density and are not usually found until later, if at all). The generated Q peaks are named as shown in the Figure 4.69 and this is observed by running the plotting program PLATON using X.res output.

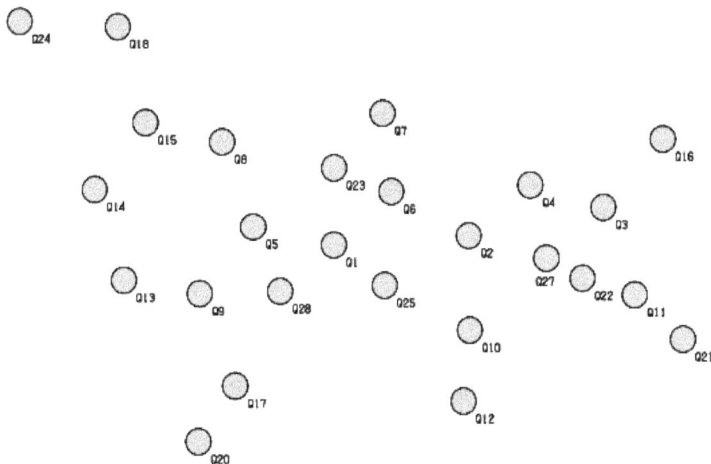

Figure 4.69 Sample E-map

In the bond geometry listed below, let us consider the first atom peak, the peak height is given as 257 and is connected to fifth peak and its bond distance is 1.419 Å and it is bonded to sixth peak and its bond distance is 1.343 Å and the bond angle (between bonds 1–5 and 1–6) is 129.9°. The remaining peaks connected to the first peak are not correct which can be known from their unusual bond lengths

Atom	Peak	x	y	z	SOF	Height	Distances and Angles					
1	257.	0.0898	0.0852	0.3931	1.000	0.37	0 5	1.419				
							0 6	1.343	129.9			
							0 19	1.606	29.8	102.6		
							0 23	1.197	80.4	49.6	55.6	
							0 25	1.067	137.7	85.7	165.6	127.5
							0 28	1.169	49.1	173.5	74.7	129.2 98.3
2	250.	0.0235	0.2427	0.5185	1.000	0.22	0 4	1.346				
							0 6	1.515	117.6			
							0 10	1.464	126.6	115.8		
							0 25	2.019	159.3	53.4	65.0	
							0 27	1.414	49.9	167.4	76.8	138.7
3	237.	0.0084	0.3900	0.6621	1.000	0.11	0 4	1.325				
							0 11	1.484	125.6			
							0 16	1.466	119.2	114.9		
							0 22	1.968	71.4	66.3	136.0	
							0 27	1.279	53.2	72.6	172.1	43.0
4	236.	0.0624	0.2993	0.6193	1.000	0.10	0 2	1.346				
							0 3	1.325	129.5			
							0 22	1.992	72.6	69.5		
							0 27	1.166	68.1	61.4	41.1	
5	235.	0.1158	0.0081	0.3384	1.000	0.38	0 1	1.419				
							0 8	1.405	121.7			
							0 9	1.401	120.9	117.4		
							0 19	0.830	99.0	40.2	134.6	
							0 23	1.697	44.1	77.6	164.9	56.1
							0 28	1.098	53.5	173.2	67.5	133.1 97.4

Sample X.1st output of SHELXS program

and bond angles. The connected peak numbers are lesser than or equal to the total number of atoms located by SHELXS program—in this case 18 only. The second peak is connected to the fourth and its bond distance is 1.346 Å and to 6th by bond distance 1.515 Å and bond angle (4–2–6) is 117.6 Å. Likewise the peak 2 is connected to peak 10 by bond distance 1.464 Å and the bond angle 4–2–10 = 126.6° and 6–2–10 = 115.8°. The connectivity diagram is plotted by the program PLATON which is shown in the Figure 4.70.

The extra Q peaks are deleted with the knowledge of the bond lengths and bond angles and with the help of the chemical diagram, and the structure is now known. Very rarely it may be a partial structure also (incomplete model). This partial structure serves as our initial model or trial structure. Usually the numbers above the total number of atoms found by SHELXS program is omitted during connectivity. If the atoms of the model structure are approximately at the right positions, there should be at least some degree of resemblance between the calculated diffracted pattern and the observed one, i.e., between the sets of $|F_c|$ and $|F_o|$ values. The two sets of values can be compared in

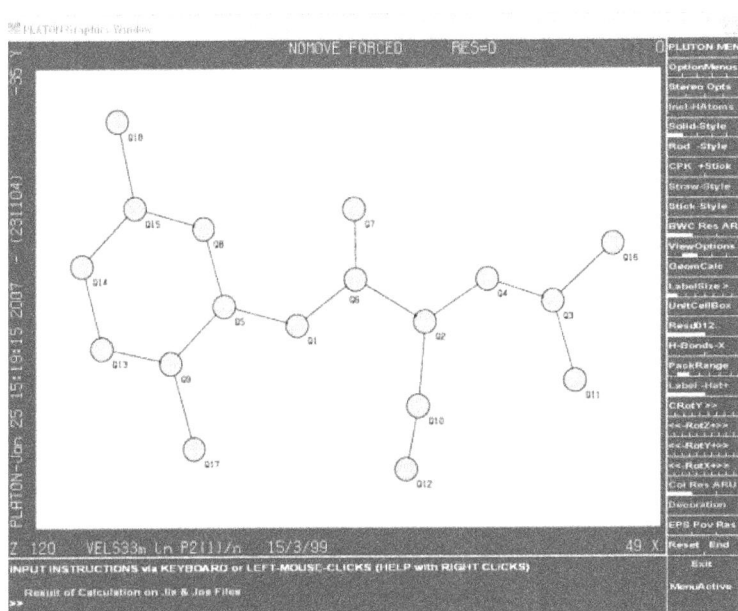

Figure 4.70 Connectivity using PLATON

various ways. The most widely used assessment is a so-called residual factor or R-factor (reliability index), defined as

$$R = \frac{\sum \left\| F_o \right| - \left| F_c \right\|}{\sum \left| F_o \right|} \qquad (4.80)$$

Since the direct methods use E-values, the R-factor in SHELXS is calculated based on E-values (in this case, $R_E = 20.7\%$). This residual factor involves adding together all the discrepancies between corresponding observed and calculated amplitudes, ignoring signs of the differences, and normalizing the sum by dividing the sum of all the observed amplitudes to give a value which can be compared for different structures. Variations on this definition include using F^2 values instead of $|F|$ values, squaring the differences, and/or incorporating different weighting factors multiplying different reflections, based on their s.u's, and hence incorporating information on the relative reliability of different measurements; for example, one residual factor in a very widely used computer program for crystal structure determination is

$$wR_2 = \sqrt{\frac{\sum w(F_o^2 - F_c^2)^2}{\sum w(F_o^2)^2}} \qquad (4.81)$$

where, each reflection has its own weight w. This is, in many ways, and certainly from a statistical viewpoint, more meaningful than the basic R factor. For a correct and complete crystal structure determined from well-measured data, R is typically around 0.03–0.07; for an initial model structure it will be much higher, possibly 0.2–0.3 depending on the fraction of electron density so far found, and its decrease during the next stages is a measure of progress. Values of wR_2 and other residual factors based on F^2 are generally higher than those based on F values, by a factor of two or more. The reverse Fourier transform carried out with the calculated amplitudes $|F_c|$ and calculated phases ϕ_c would just regenerate the electron density of the model structure and this is not the progress. However, combination of the experimentally observed amplitudes $|F_o|$ (which carry information about the true structure) with the calculated phases ϕ_c (which are not completely correct, but are the best approximation we currently have to the unavailable ϕ_o values) produces new model structure. Usually, if the errors in the calculated phases are not too large, this electron density shows the atoms of the existing model structure, together with additional atoms not already known. This provides an improved model structure, with more atoms than before.

If there are still more atoms to be found, this process can be repeated. A forward Fourier transform of the new model structure gives a new set of $|F_c|$ and ϕ_c; the previous set is discarded. The new $|F_c|$ and the unchanged original $|F_o|$ values should now give a lower R-factor, and the improved ϕ_c together with $|F_o|$ generate—via another reverse Fourier transform—a further electron density map. Eventually all the atoms are located and the Fourier transform calculations give no further improvement. The reverse Fourier transform can be carried out using the differences $|F_o| - |F_c|$ instead of $|F_o|$ and the difference electron density

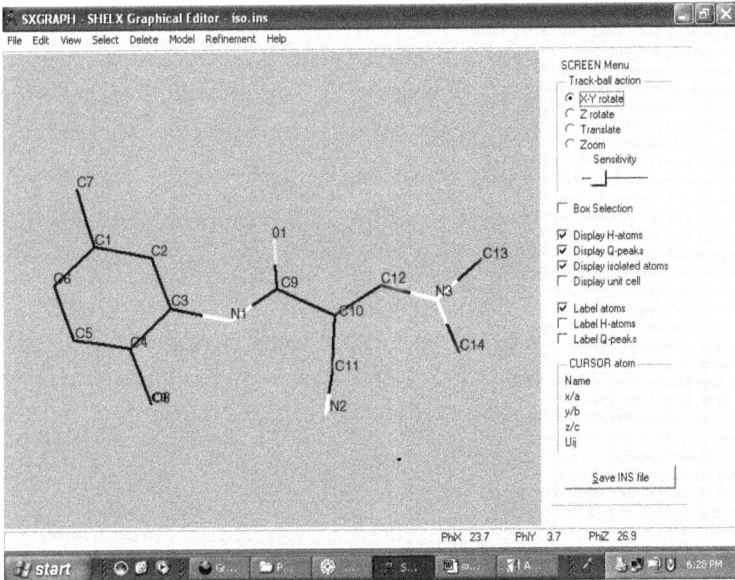

Figure 4.71   Model structure

map is created in which the existing atoms of the current model structure do not appear. This makes new atoms stand out more clearly from the background and from false maxima arising from errors in the ϕ_c values. Difference electron density peaks or holes (negative peaks) at model structure atom positions may indicate incorrect atom assignments with too little or too much assumed electron density, which should be corrected in the next model and the resulted model structure of the example compound, is shown in Figure 4.71.

Structure Refinement

Once all the non-hydrogen atoms have been found, the model structure needs to be refined. This means varying the numerical parameters describing the structure to produce the 'best' agreement between the diffraction pattern calculated from it by a Fourier transform and the observed diffraction pattern. Since there are no observed phases, the comparison of observed and calculated diffraction patterns is made entirely on their amplitudes $|F_o|$ and $|F_c|$. Changing any of the structural parameters affects the $|F_c|$ values, while the $|F_o|$ values remain fixed during the process.

The refinement process uses a well-established mathematical procedure called **least squares analysis**, which defines the "best fit" of two sets of data to be that which minimizes one of the least squares sums

$$\sum w(|F_o| - |F_c|)^2$$

or

$$\sum w(F_o^2 - F_c^2)^2$$

The first of these (refinement on F) has historically been more commonly used, but the second (refinement on F^2) is now increasing popularity and is, in many ways, superior. The contribution of each reflection to the sum is weighted according to its perceived reliability, usually with weights based on the experimental s.u.'s, such as $w = \frac{1}{\sigma^2(F_o^2)}$ for refinement on F^2. The parameters to be refined are the positions and vibrations of the atoms in the Fourier transform equation. For each atom, there are three positional coordinates x, y, z and a displacement parameter U, which can be interpreted as an *isotropic mean-square amplitude of vibration* (in Å^2) of the atom. A significantly better fit to the data can be achieved by using more than one displacement parameter per atom in the model structure, allowing each atom to vibrate by different amounts in different directions (*anisotropic vibration*). This will be done by adding ANIS command in the instruction file and by refining using SHELXL 97. The usual mathematical treatment has six U values (one for each axis and three cross-terms) for each atom in order to give different vibration amplitudes in three orthogonal directions which are, in general, not along the unit cell axes. Thus there are commonly nine refined parameters (three for position and six for thermal parameter) for each independent atom (atoms which are not related to each other by symmetry) in the structure. The scale factor is an additional parameter refined.

The least square process provides value for each refined parameters and also gives a standard uncertainty (s.u.). These s.u.'s depend on the s.u.'s of the data, on the extent of agreement of the observed and calculated data (a lower least squares sum gives lower parameter s.u.'s; another function closely related to this sum is called the goodness of fit), and on the excess of data over parameters (a greater excess gives a lower parameter s.u.'s). Both the quality and the quantity of measured data influence the quality (reliability) of the model derived from them.

Once the model structure has been refined with anisotropic displacement parameters for the atoms, it is often possible to see small but significant difference electron density peaks in positions close to those expected for hydrogen atoms, particularly if there are few or no heavy atoms in the structure. Hydrogen atoms are more likely to be located from measurements taken at low temperature, because this reduces the vibration of the atoms and sharpens the electron density peaks. It is possible to include the hydrogen atoms in the refinement by geometrically fixing them. This may improve the fit slightly but their parameters usually have large s.u.'s because, their low electron density contributes a mean for hydrogen atoms only weakly to the diffraction of X-rays. In most cases, refinement is more

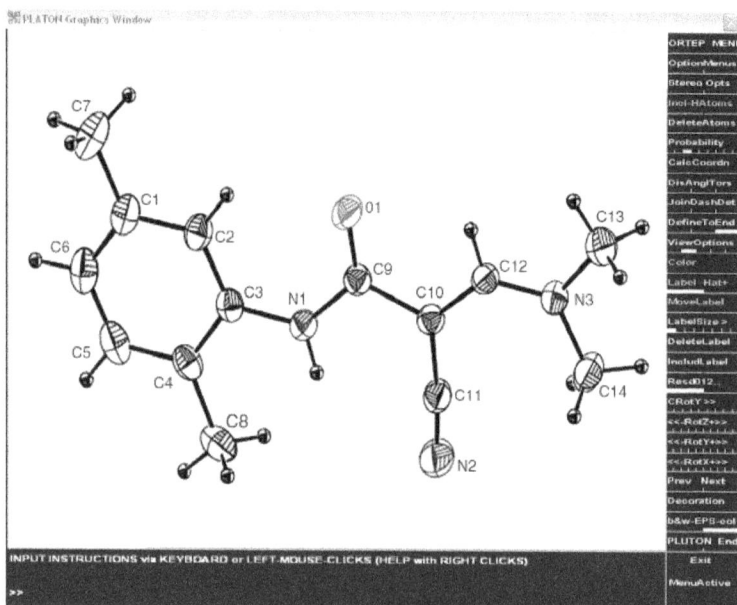

Figure 4.72   Thermal ellipsoidal plot

successful if constraints are applied to hydrogen atom parameters, for example, by keeping the bond lengths fixed and by assigining the U values to those of the atoms to which they are bonded. After anisotropic refinement and location of hydrogen atoms, the thermal ellipsoidal plot (using ORTEP program) is drawn for the example compound and this is shown in Figure 4.72. For publication purposes, ACTA, CONF and BOND $H commands are used to create crystallographic information file (CIF) with the bond geometry tables.

At the end of refinement, a difference electron density map, should not contain any significant features (peaks or holes). This calculation is usually performed as an extra check on the validity of the refined model structure. Typically, a final map with no features outside the range ± 1 eÅ^{-3} is accepted as an evidence of a satisfactory structure determination.

Structural Analysis

The outcome of the crystal structure analysis are the unit cell geometry and symmetry (space group), and the positions of all the atoms in the unit cell (three coordinates each), together with their isotropic (one) or anisotropic (six) displacement parameters. The displacement parameters are usually interpreted as representing thermal vibration of the atoms. From the atomic coordinates, unit cell geometry and symmetry, many geometrical results can be derived. These include:

Bond lengths, bond angles and torsion angles   The bond lengths are calculated by the program as follows:

Let (x, y, z) be the fractional coordinates of an atom in the crystallographic coordinate system and (x', y', z') be the coordinates of an atom in the crystallographic coordinate system in Å units and (X, Y, Z) be the coordinates of the atom in orthogonal coordinate system. The fractional coordinates are transformed into coordinates in Å by $x' = ax$, $y' = by$, $z' = cz$ where a, b, c are the cell parameters. For the transformation to orthogonal coordinate system, the relation is

$$\begin{pmatrix} X \\ Y \\ Z \end{pmatrix} = T \begin{pmatrix} x' \\ y' \\ z' \end{pmatrix}$$

where, $T = \begin{pmatrix} 1 & \cos\gamma & \cos\beta \\ 0 & \sin\gamma & P \\ 0 & 0 & \sqrt{\sin^2\beta - P^2} \end{pmatrix}$

where, $P = \dfrac{\cos\alpha - \cos\gamma\cos\beta}{\sin\gamma}$

Let us consider two atoms at $A(x_A, y_A, z_A)$ and $B(x_B, y_B, z_B)$ with position vectors $\vec{r}_A$ and $\vec{r}_B$.

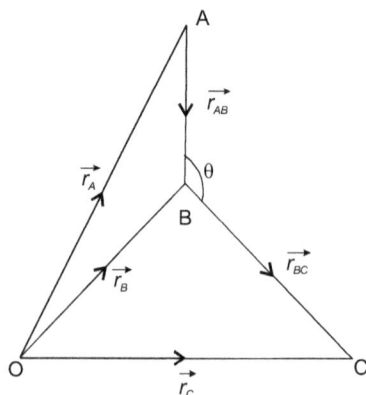

$$\overrightarrow{AB} = \overrightarrow{r}_{AB} = \overrightarrow{r}_{B} - \overrightarrow{r}_{A}$$

$$\overrightarrow{AB} = \left|\overrightarrow{r}_{AB}\right| = \left|\overrightarrow{r}_{B} - \overrightarrow{r}_{A}\right|$$

$$= \left|(x_{B} - x_{A})\vec{i} + (y_{B} - y_{A})\vec{j} + (z_{B} - z_{A})\vec{k}\right|$$

Therefore, bond length between the two atoms

$$AB = [(x_{B} - x_{A})^{2} + (y_{B} - y_{A})^{2} + (z_{B} - z_{A})^{2}]^{1/2}$$

Similarly the bond angles are calculated as follows.

Let us consider three atoms at A, B, and C with position vectors $\overrightarrow{r}_{A}$, $\overrightarrow{r}_{B}$, and $\overrightarrow{r}_{C}$. The bond angle $\theta = \angle ABC$ can be calculated as follows.

$$\overrightarrow{BA} = \overrightarrow{r}_{A} - \overrightarrow{r}_{B} = (x_{A} - x_{B})\vec{i} + (y_{A} - y_{B})\vec{j} + (z_{A} - z_{B})\vec{k}$$

$$\overrightarrow{BC} = \overrightarrow{r}_{C} - \overrightarrow{r}_{B} = (x_{C} - x_{B})\vec{i} + (y_{C} - y_{B})\vec{j} + (z_{C} - z_{B})\vec{k}$$

$$\overrightarrow{BA} \cdot \overrightarrow{BC} = (x_{A} - x_{B})(x_{C} - x_{B}) + (y_{A} - y_{B})(y_{C} - y_{B}) +$$
$$(z_{A} - z_{B})(z_{C} - z_{B})$$
$$= N$$

Also, $\overrightarrow{BA} \cdot \overrightarrow{BC} = \left|\overrightarrow{AB}\right|\left|\overrightarrow{BC}\right|\cos\theta = D\cos\theta$

where, D = length AB × length AC

Therefore $\cos\theta = \dfrac{N}{D}$

$$\theta = \cos^{-1}\left(\dfrac{N}{D}\right)$$

The torsion angle is calculated as

$$\theta = \cos^{-1}\left(\frac{\overrightarrow{N_1}.\overrightarrow{N_2}}{|\overrightarrow{N_1}||\overrightarrow{N_2}|}\right)$$

where, N_1 and N_2 are the unit normalized vectors to plane 1 and plane 2. The sign of the angle is determined by the vector products

$$\text{sign} = (\overrightarrow{N_1} \times \overrightarrow{N_2}) \cdot \overrightarrow{BC}$$

where, BC is the common vector.

Thus the bond lengths, bond angles and torsion angles are determined by their position coordinates and are listed in the output file after final refinement.

Conformations of rings   **Conformation** is defined as the spatial arrangement of the atoms which can be interconverted by rotations about formal single bonds. Conformational change from the planar arrangement of atoms of a five-membered or six-membered ring may be due to some bulky groups attached to that particular ring and its steric hindrance. Steric effects arise from the fact that each atom within a molecule occupies a certain amount of space. If atoms are brought too close together, there is an associated cost in energy due to overlapping electron clouds (Pauli or Born repulsion), and this may affect the molecule's preferred shape (conformation) and reactivity.

Program called PARST is used to calculate the asymmetry parameters from which the conformations of the rings are interpreted. The five-membered and six-membered rings adopt any one of the following possible conformations which are shown in Figures 4.73 and 4.74 respectively.

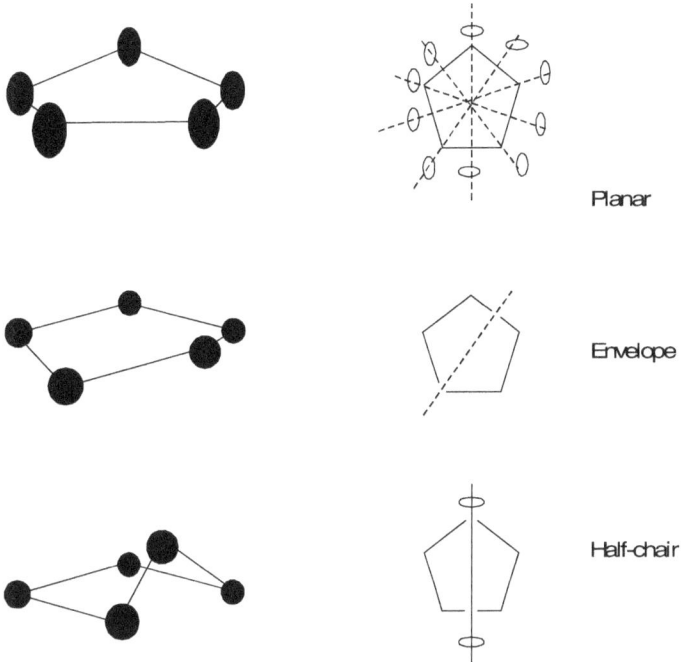

Planar

Envelope

Half-chair

$$\Delta C_s = \left[\frac{\displaystyle\sum_{i=1}^{m}\left(\phi_i + \phi_i'\right)^2}{m} \right]^{1/2}$$

$$\Delta C_2 = \left[\frac{\displaystyle\sum_{i=1}^{m}\left(\phi_i - \phi_i'\right)^2}{m} \right]^{1/2}$$

Mirror asymmetry   Two-fold asymmetry

where, ϕ_i, ϕ_i' are symmetry related torsions

m is the number of individual comparisons

Figure 4.73 Possible confirmations of a five-membered ring

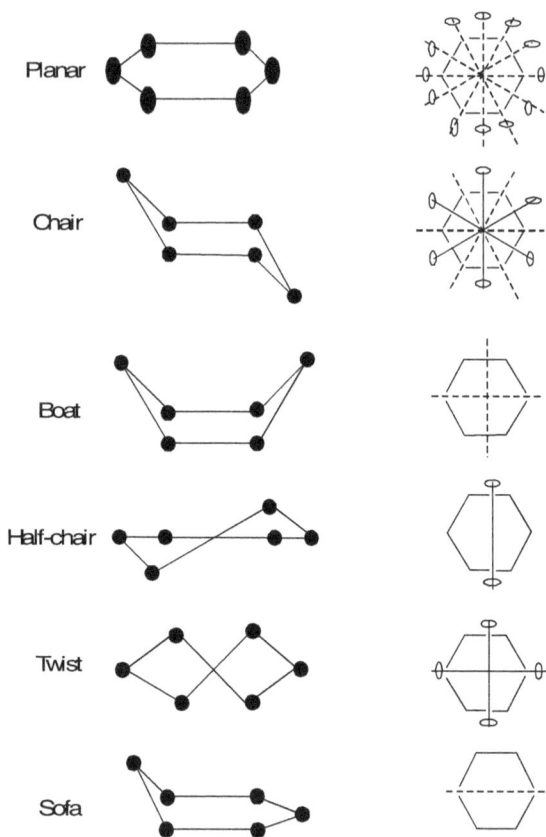

$$\Delta C_s = \left[\frac{\sum\limits_{i=1}^{m} \left(\phi_i + \phi_i' \right)^2}{m} \right]^{1/2} \qquad \Delta C_2 = \left[\frac{\sum\limits_{i=1}^{m} \left(\phi_i - \phi_i' \right)^2}{m} \right]^{1/2}$$

Mirror asymmetry   Two-fold asymmetry

where, ϕ_i, ϕ_i' are symmetry related torsions

m is the number of individual comparisons

Figure 4.74   Possible conformations of a six-membered ring

Planarity Plane calculations were carried out by PARST and the puckering distances are calculated and from the results, the deviations of substituent atoms from the ring to which it is attached are known.

Molecular packing Molecular packing can be visualized by the software called **PLATON** in various pack ranges. The molecular packing diagram of the example compound is shown in Figure 4.75.

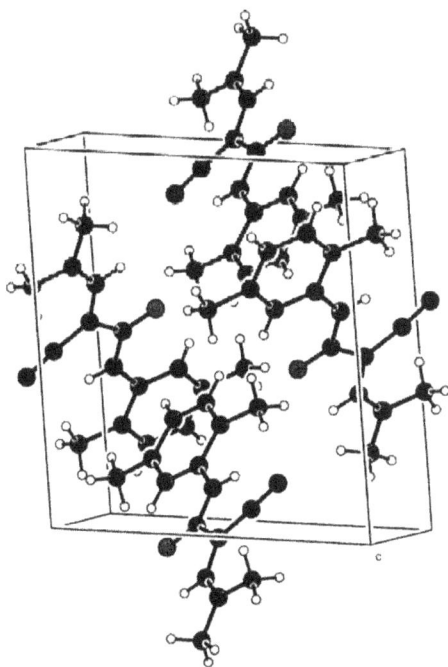

Figure 4.75 Molecular packing diagram of example compound

Inter- and intramolecular hydrogen bonds Hydrogen bonding is the specific type of non-bonded interaction between two electronegative atoms, where hydrogen atom is covalently bonded to one of them. The usual way of representing the hydrogen bond is D–H...A, where D is the donor and A is an acceptor. The hydrogen bonds are highly directional and the D–H...A angle should be 180° for an ideal one. The distance between the donor and acceptor atom is shorter by at least 0.2 Å than the sum of their van der Waals radii for the hydrogen bond to be present in the crystal structure. The proton donor capacity of a C–H group depends on the hybridization [Csp–H > Csp^2–H > Csp^3–H] and increases with the number of adjacent withdrawing groups. The strong and weak hydrogen bonds are listed out using CALC ALL option in PLATON graphical menu. The intra- and intermolecular hydrogen bonds (N–H...N and C–H...N) for the example structure is shown in Figure 4.76.

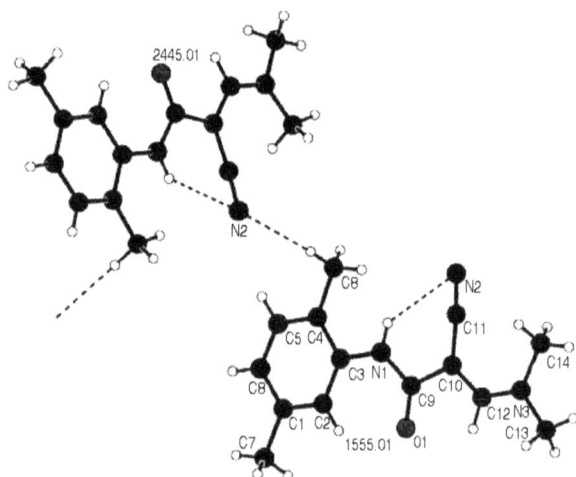

Figure 4.76 Inter- and intramolecular hydrogen bonds

Non-bonded interactions (C–H...π) The molecular packing is also stabilized by some non-bonded interactions. In this case, C–H...π interaction involves C15 atom, the hydrogen bonded to C15 and the centre of the aromatic ring of the symmetry related molecule as shown in Figure 4.77.

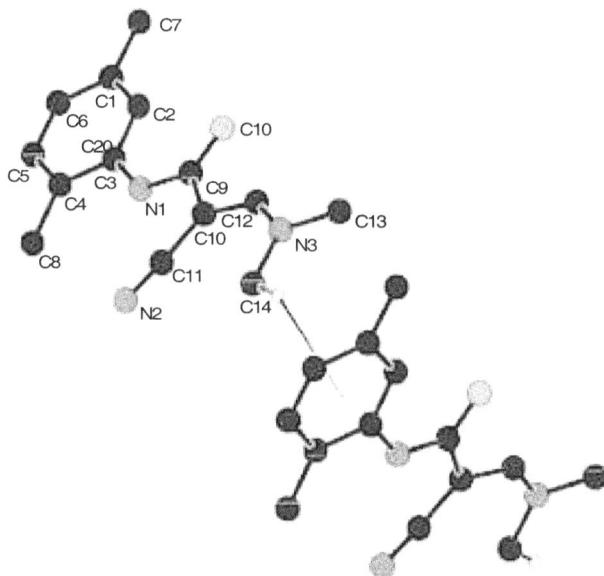

Figure 4.77 C–H...π interaction

The non-bonded interactions in some of the structures are depicted here.

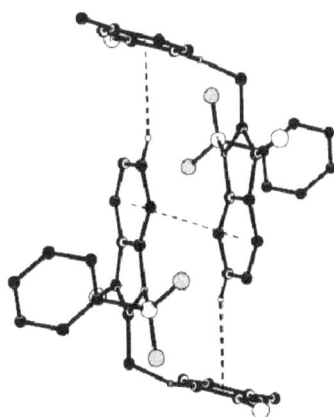

C–H...π and π...π interactions

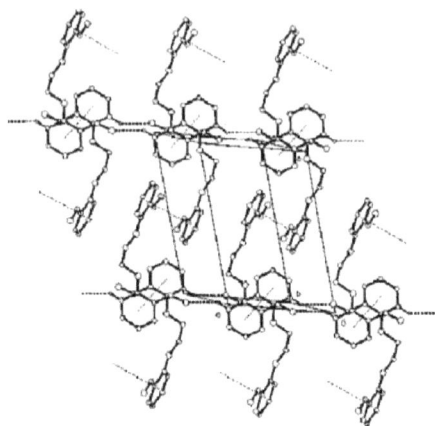

C-H...O and $\pi...\pi$ interactions

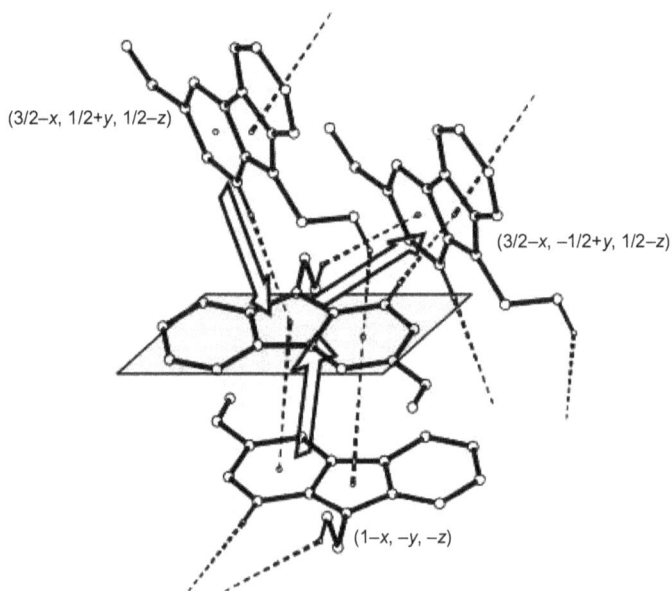

$(3/2-x, 1/2+y, 1/2-z)$

$(3/2-x, -1/2+y, 1/2-z)$

$(1-x, -y, -z)$

$\pi...\pi$ and C–H...π interactions

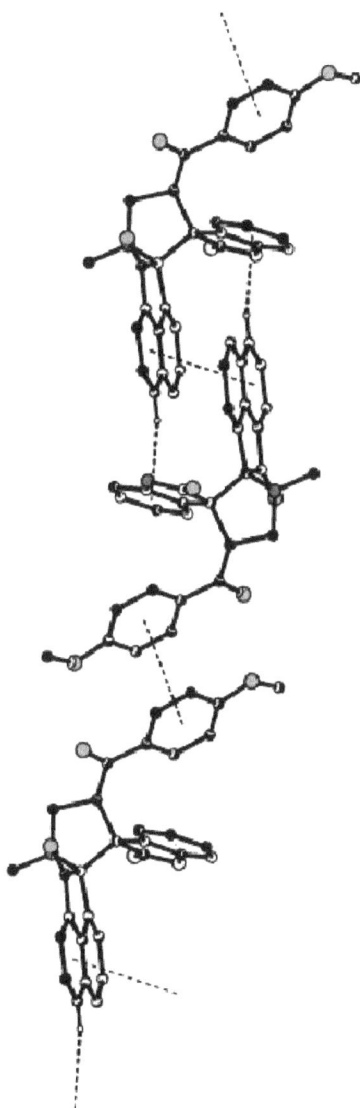

C–H...π and π...π interactions

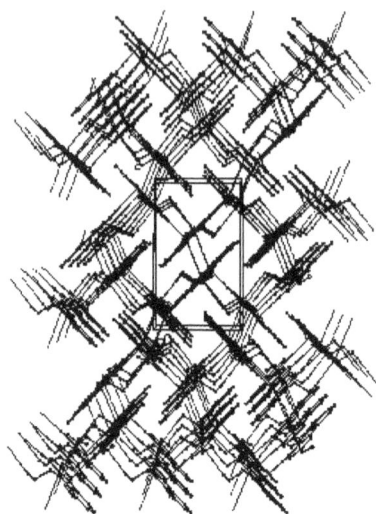

Herring-bone packing (C–H...π and π...π interactions)

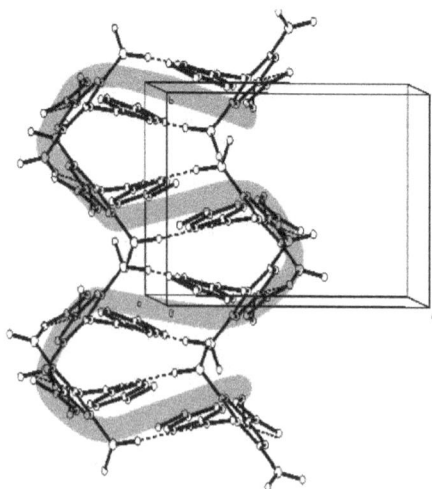

Ladder type packing with hydrogen bonds

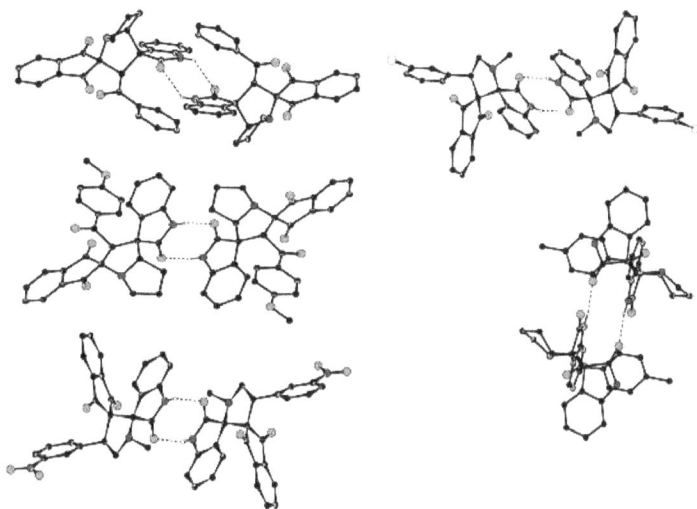

N–H...O hydrogen-bonded dimer

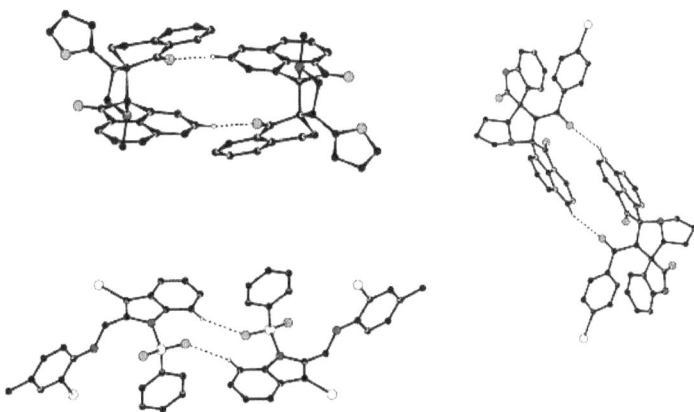

C–H...O hydrogen-bonded dimer

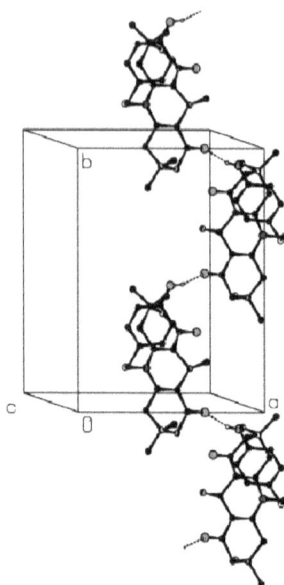

O–H...O hydrogen-bonded molecular chain

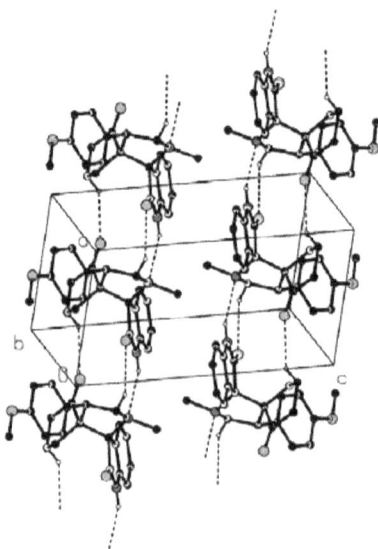

N–H...O and C–H...O hydrogen-bonded molecular ribbons running along *a*-axis

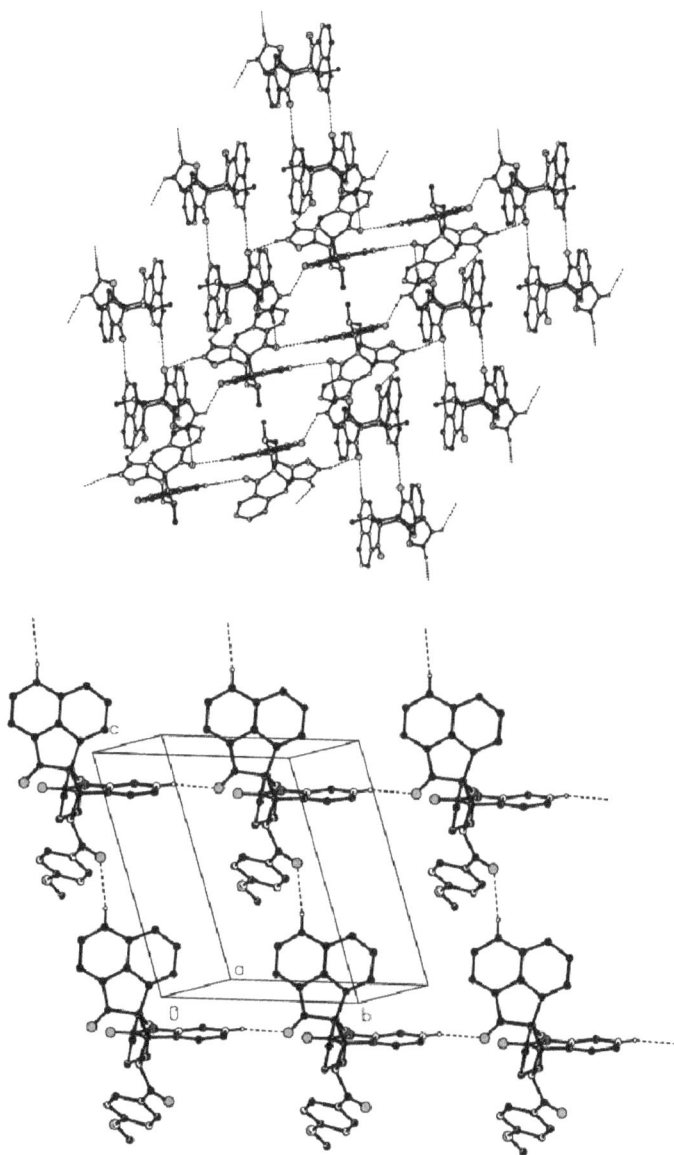

Molecular sheets

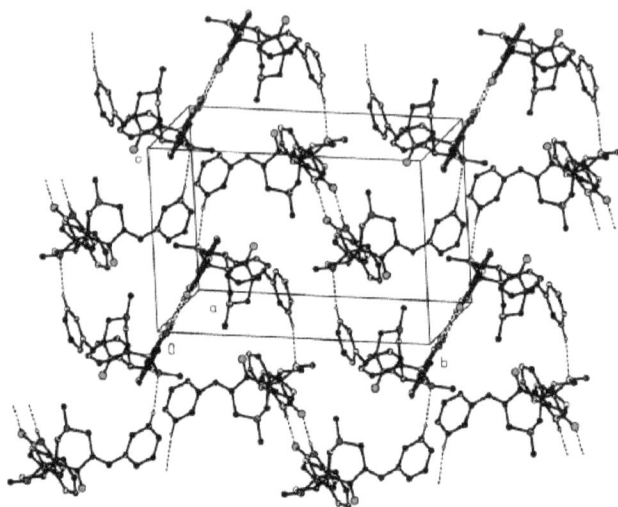

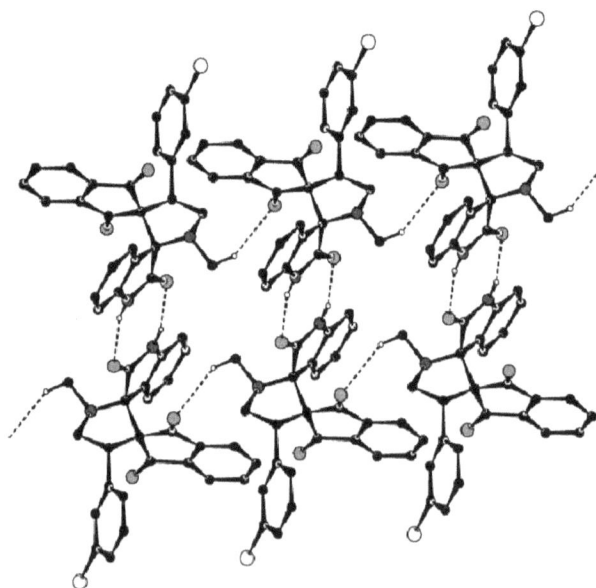

Molecular sheets

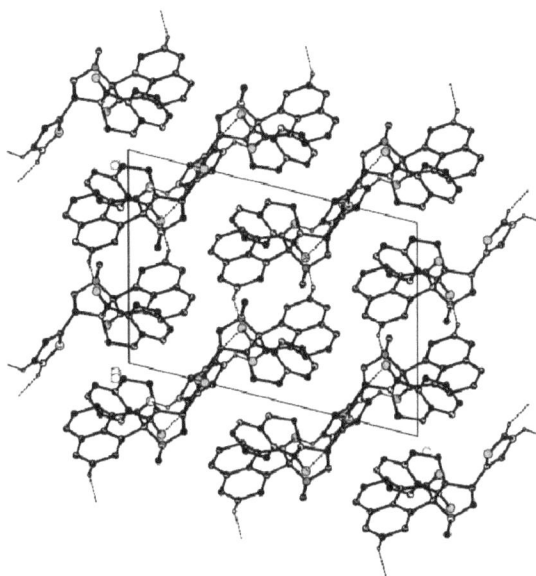

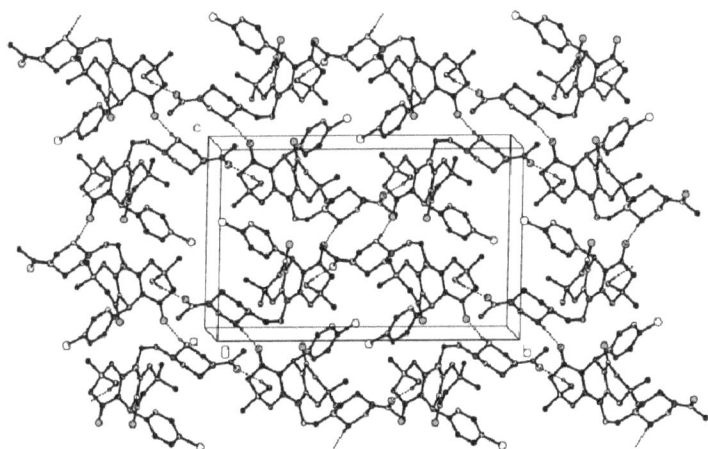

Molecular packing

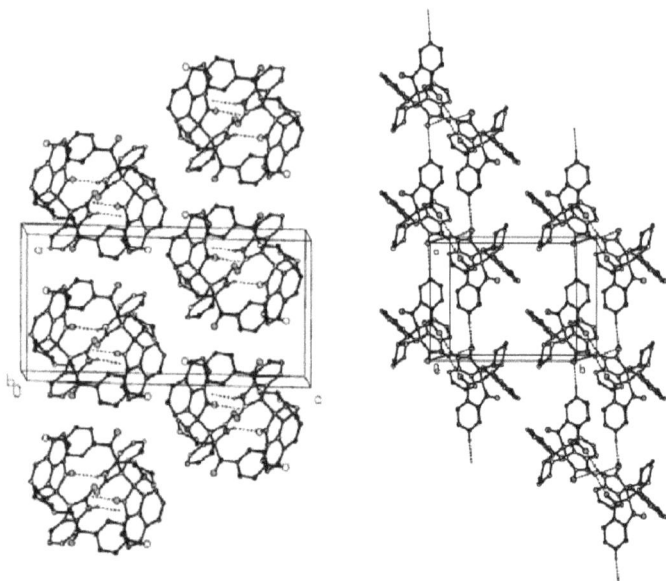

Crystal packing

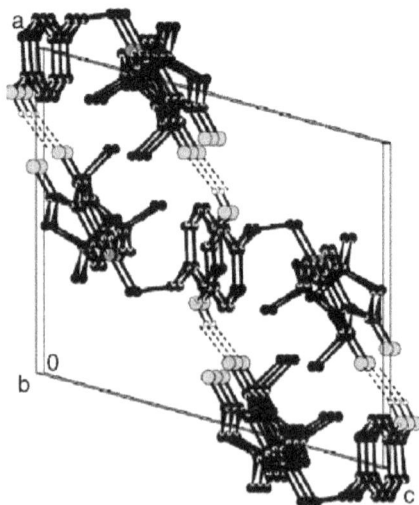

Molecular assembly

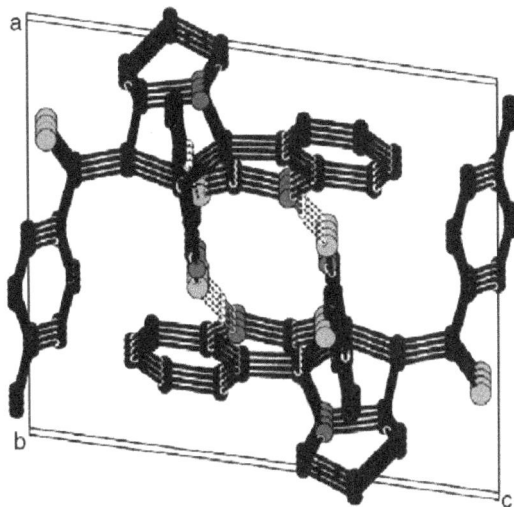

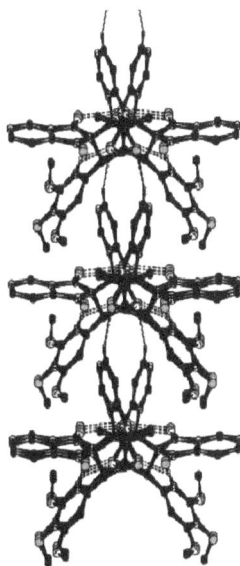

Molecular assembly

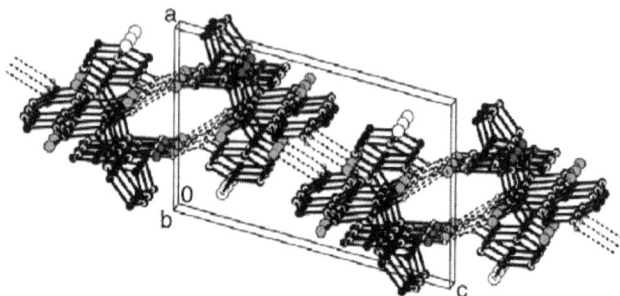

Molecular assembly viewed down *b*-axis

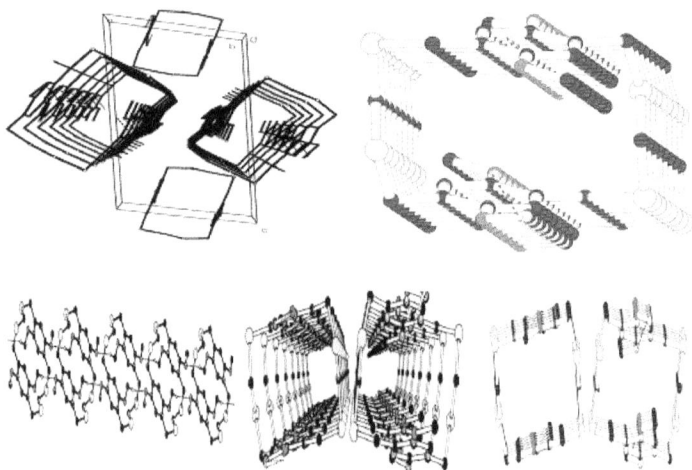

Tunnel-like patterns

ELECTRON DIFFRACTION

The wave nature of electrons is confirmed by de Broglie postulates. The beam of electrons incident on a crystal suffer Bragg's diffraction like X-rays. Electron crystallography opens up new possibilities for the crystal structure determination of biological macromolecules. Electron crystallography works with well-ordered two-dimensional crystals of proteins that require less material, and these crystals form more readily than single three-dimensional crystals. Thus the technique of electron crystallography appears particularly suited for complex membrane proteins, since it helps to overcome the main difficulties in the crystallization of these proteins. In addition, electron crystallography can be applied to smaller samples than X-ray diffraction due to the strong interaction of electrons with matter.

Electron crystallography is a method to determine the arrangement of atoms in solids using an electron microscope. It can complement X-ray crystallography on proteins (such as membrane proteins), that cannot easily form the large three-dimensional crystals required for that process. Structures are usually determined from either two-dimensional crystals (sheets or helices), polyhedrons such as viral capsids, or dispersed individual proteins. Electrons can be used in these situations, whereas X-rays cannot, because electrons interact more strongly with atoms than X-rays do. Thus, X-rays will travel through a thin two-dimensional crystal without diffracting significantly, whereas electrons can be used to form an image. Conversely, the strong interaction between electrons and proteins makes thick (e.g. three-dimensional) crystals impervious to electrons, which only penetrate short distances.

The electron, like the neutron, possesses wave properties, and the wavelength according to quantum mechanics is given by

$$\lambda = {h}/{p} \qquad (4.82)$$

where, p is the momentum of the particle and h is the Planck's constant.

Consider an electron, which is initially at rest where the electrostatic potential is zero, moves to a point where the potential is P. Equating the reduction in potential energy to the gain in kinetic energy,

$$Pe = \frac{P^2}{2m} \qquad (4.83)$$

where, e is the charge of electron and m is the mass of the electron. Inserting the values of the constants h, m, and e, we get

$$\lambda = \sqrt{\frac{1500}{P}} \qquad (4.84)$$

LOW-ENERGY ELECTRON DIFFRACTION

Davisson and Germer performed the diffraction experiment with low-energy electrons (30–600 eV). The diffraction effects of a crystal with such low-energy electrons are due to a very few layers of atoms, next to the surface. Consider a row of atoms normal to the plane of paper and lying in the surface of a crystal (Figure 4.78).

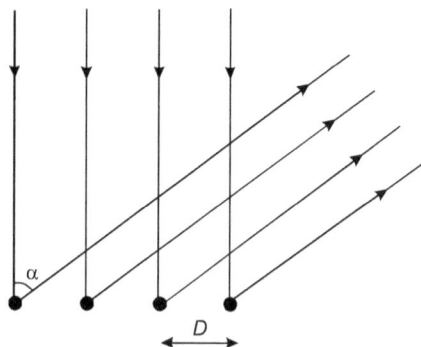

Figure 4.78   Low-energy electron diffraction

Let a beam of electrons be incident normally on the surface of a crystal, and be diffracted backwards through an angle α. The waves scattered by adjacent rows of atoms will interfere to give a diffraction maximum, if the path difference is an integral number of wavelengths. The condition for the diffraction maximum is as given by Bragg's law.

$$D \sin \alpha = n\lambda, \tag{4.85}$$

where,

D is the distance between the rows of atoms,

n is an integer and

λ is given by the equation 4.84.

Davisson and Germer used electrons of wavelength 0.5–1.5 Å. Since the spacing of rows of atoms in a crystal is typically of the order of 1.5 Å, the angle of difraction α will be at least 20°.

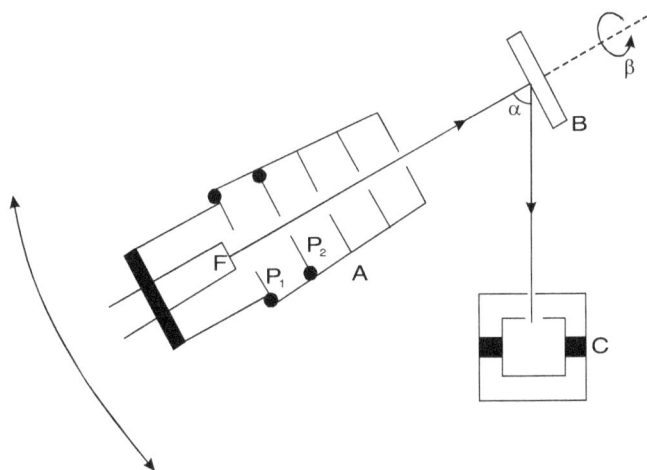

Figure 4.79 Davisson and Germer apparatus for low-energy diffraction

The apparatus (Figure 4.79) consists of electron gun A. The tungsten filament F is situated in a rectangular slot in a diaphragm P_1. The potential of this diaphragm is made slightly negative with respect to the filament. The resulting repulsive field between P_1 and the filament helps to concentrate the emitted electrons into a rather divergent beam. The electrons are then accelerated towards the diaphragm P_2, which is maintained at a fairly high positive potential, and some of them pass through a 1 mm diameter aperture to form a narrow, well collimated beam. Three further apertures, maintained at a rather lower potential than P_2, provide a further collimation of the beam. These apertures are each about 1 mm diameter and 8 mm apart. The beam of electrons from the gun strikes the (111) face of a single crystal of nickel B situated about 10 mm away from the gun. The scattered electrons are collected in the double-walled Faraday cylinder C, which has an entrance aperture 1 mm diameter, situated about 10 mm from the crystal B. The potential of the outer cylinder is the same as that of the crystal and the outer electrode of the gun. The inner cylinder is connected through a galvanometer to a point, which is at a small positive potential with respect to the filament. This ensures that only the electrons, which have been scattered without appreciable loss of energy, are recorded.

The whole electrode system is mounted in a metal box to ensure that the electrons move in a space free of electrostatic fields between the gun and the collector. This box is itself contained in a glass bulb, which can be evacuated. After a thorough baking out, the bulb is sealed off from the pumps, and it is estimated that the residual gas pressure inside it is then only about 10^{-8} Torr.

The energy of the diffracted electrons is too low to affect a photographic plate or Geiger counter, and the only simple method of detection is to collect them in a Faraday cylinder and to measure the current. The collector C is mounted on an arm, so that it can rotate about an axis perpendicular to the plane of the figure and hang, pendulum wise, vertically below B. By rotating the evacuated bulb and its contents about this axis, the direction of the scattered electrons entering the collector can be varied. The pendulum-like system thus makes it possible by rocking the apparatus to rotate the crystal about an axis normal to its surface and by coinciding with the incident electron beam from the gun. The distribution of the scattered electrons can thus be explored both in co-latitude (α) and azimuth (β).

If electrons of higher energy are used with P > 100 volts, then the diffraction effects are more only at narrow range of P. This means that the electrons are able to penetrate into the crystal a few atom layers, and diffraction occurs from the successive layers of atoms. Thus when P is varied, the angle α of the scattered beam is found to vary according to the equation

$$D \sin \alpha = \sqrt{1500/P} \qquad (4.86)$$

The periodic potential of the lattice terminates at the crystal surface; the arrangement of the atoms on the surface may differ from that in the bulk. The structure of the thin surface layer may not be the same with that of the remaining part of the crystal. Also the crystal surface plays an important role in electron and ion emission, adsorption and catalysis, nucleation of a new phase and diffusion, ionic implantation, oxidation, etc. Thus low-energy electron diffraction is an effective method for investigating the crystal surface for the arrangement of atoms on it, the nature of their thermal vibrations, etc.

HIGH-ENERGY ELECTRON DIFFRACTION

Thomson used values of potential P in the range 10–60 kV in high-energy diffraction experiment. The electrons with this energy can penetrate solid films several hundred atom layers in thickness without any appreciable loss of energy. The wavelength of such electrons is less than 10 pm, and the angle between the incident and diffracted beams is therefore only a few degrees. The high-energy electron diffraction apparatus is shown in Figure 4.80.

A cold cathode discharge tube A at low pressure is the source of electrons. The induction coil provides the necessary high potential and is measured by a spark-gap voltmeter.

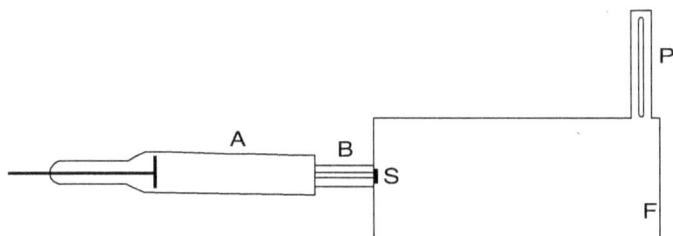

Figure 4.80   High-energy electron diffraction apparatus

The electrons pass through a hole in the anode and are collimated by a narrow tube B, about ¼ mm diameter and 60 mm long. The collimator also serves to reduce the leakage of gas from the discharge tube into the part of the apparatus to the right, where a pressure of 10^{-8} Torr is maintained. The collimator tube is surrounded by a thick iron tube. This serves to screen off the Earth's magnetic field, which would therefore cause the electrons to move in a curved trajectory and therefore be unable to pass through the collimating tube. On emerging from the collimator, the beam passes through the specimen S in the form of a thin film. At a distance of 325 mm beyond S, the electrons strike a fluorescent screen F, on which the diffraction pattern can be observed. A photographic plate P can be lowered in front of the screen to have a permanent record.

The electron diffraction work is carried out mostly by electron microscope nowadays. An electron microscope can produce excellent diffraction patterns, and the interpretation of these is often helped by the ability to correlate them with electron micrographs of the specimens. Also, it is possible to magnify the diffraction pattern by the electron lenses of the microscope so that fine details become easily visible. With a conventional electron microscope, it is easily possible to produce a diffraction pattern of a size corresponding to a simple camera some tens of metres in length. The modern electron microscope has five magnetic lenses.

COMPARISON OF LOW-ENERGY AND HIGH-ENERGY ELECTRON DIFFRACTION

The low-energy electrons are diffracted mainly by the surface layer of atoms and a high vacuum is required. As a result of the greater strength of the diffraction of electrons, it is possible to determine a crystal structure using less than 10^{-10} g of material using high-energy diffraction.

ELECTRON DIFFRACTION CAMERA

The electron diffraction camera consists of a magnetic focusing lens, which was improved by Lebedeff. In the earliest cameras, this consisted of a short coil of wire with its axis roughly parallel to the direction of the electron beam; a steady current producing an excitation of about 2000 ampere-turns is passed through the coil. The effect of such a coil on the electron trajectories is very similar to that of a converging lens on rays of light. Figure 4.81 shows the electron trajectories in such a simple diffraction camera. The magnetic lens is represented schematically at L as if it were an optical lens. Electron beams from a source G are focused by L on to a photographic plate or fluorescent screen. S is the specimen holder and the undeviated beam is focused at C. The waves diffracted from the atomic planes are deviated through the angles CAD and CBD each equal to 2θ. From the geometry of the circle, it is known that the points ABCD must lie on a circle which means that the various diffracted beams are focused on a sphere of diameter approximately equal to the specimen-plate distance. The replacement of the two fine collimating apertures of Thomson's original apparatus by a single aperture and a magnetic lens results in both a more sharply defined and a more intense diffraction pattern.

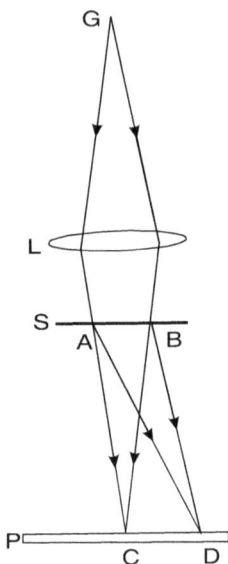

Figure 4.81 Optical system in simple high-energy camera

Simple Electron Diffraction Camera and its Working Principle

The general arrangement of the camera is shown in Figure 4.82 in which the camera body is usually made of iron or steel, as it is good for vacuum engineering. The electrons are produced by the electron gun G. The electron beam is focused by the magnetic field supplied by the beam aligning magnets M. The magnetic focusing lens L serves to image the source of electrons on to the photographic plate P. D is the diaphragm to focus the beam on the specimen holder S. The vacuum is created by using the vacuum port R on specimen chamber C. The diffracted electron beams are recorded on the fluorescent screen F and by photographic plate P.

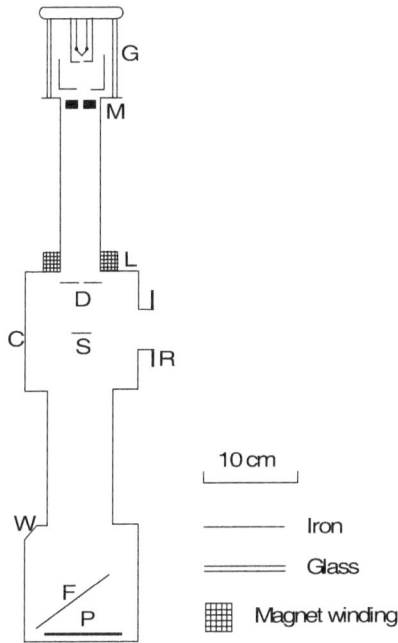

Figure 4.82 Simple electron diffraction camera G—Electron gun;
M—Beam aligning magnets; L—Magnetic focusing lens;
C—Specimen chamber; D—Diaphragm; S—Specimen;
R—Vaccum port on specimen chamber; F—Fluorescent screen
shown partly raised; P—Photographic plate; W—Window

Electron gun Electron gun, which is at the top of the camera
Figure 4.83, serves as the source for generating electrons.
This is more reliable and stable than the gas discharge tube.
The source of electrons is a V-shaped filament F made of
pure tungsten wire about 100 mm diameter and between 10
and 20 mm in length; this is heated by a current from the
transformer T having a secondary winding insulated to
withstand the accelerating potential (up to 100 kV) which is
applied between the filament and earth. The filament is
sharply kinked at its tip as shown. This has the effect of

slightly reducing the heat loss from this part of the filament, so that the temperature of the tip is slightly higher than that of the rest of the filament. Since the thermionic emission increases very rapidly with temperature, it follows that the bulk of the emission comes from a very small portion of the filament at its tip; in other words, we have a good approximation to a point source of electrons.

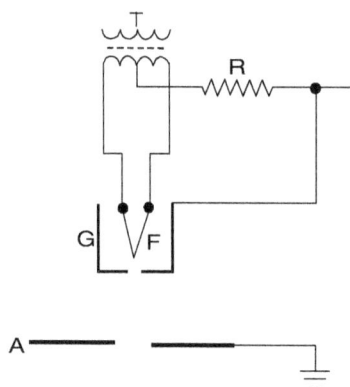

Figure 4.83   Electron gun

Magnetic focusing lens   The electron lens L (Figure 4.84) serves to image the source of electrons on to the photographic plate P or fluorescent screen F at the bottom of the camera. The general principle of its construction is shown in Figure 4.84, which also shows how a vacuum-tight connection is made between the lens and the upper part of the camera. The diametrical section of the magnetic lens is shown in the Figure 4.84. This is cylindrically symmetrical about the electron beam, which is indicated by a vertical arrow. The upper part of the camera terminates in a flange F, which is fastened to the top plate of the lens by bolts B. A rubber O-ring, compressed between the flat upper surface of the lens and a recess in F, makes a reliable vacuum-tight connection.

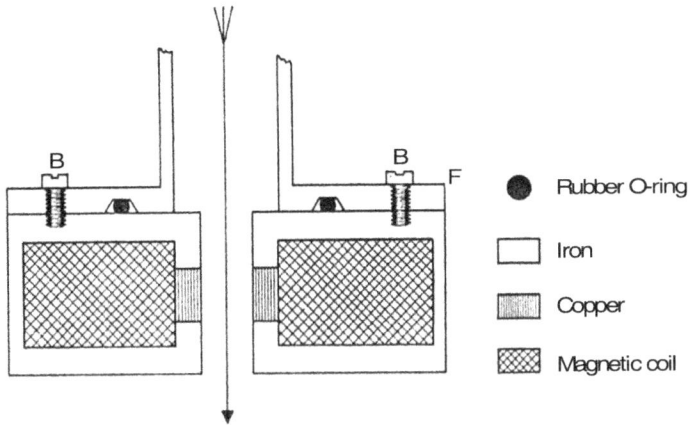

Figure 4.84 Magnetic focusing lens

Specimen chamber The specimen chamber C is situated below the magnetic focusing lens. This is often fairly large— 150 mm diameter and 150 mm long—so that auxiliary apparatus can be mounted in the vacuum, in order to measure the other physical properties of the specimen while observing its diffraction pattern. The specimen S is mounted on a special holder that is not shown in Figure 4.85 and

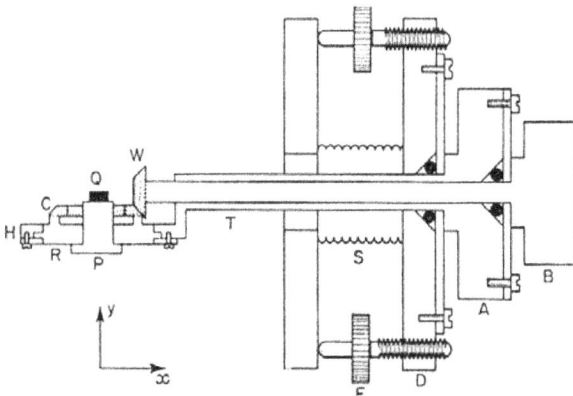

Figure 4.85 Specimen chamber

it fits into the port R. The specimen holder is in such a way that it can be adjusted from outside the vacuum system, which is made possible by rotating the knobs A and B shown in Figure 4.85.

High and low temperature specimen mounts   A miniature electric furnace is attached to the specimen holder, which enables to study specimens at high temperatures. The arrangement of a typical furnace for use with transmission specimens is shown in Figure 4.86. The specimen is placed over a cavity in a copper or silver block A. It is clamped in position by the flanged tube B. A heating coil C is wound on the block A, and the whole is surrounded by the metal radiation shield D, which is attached to the furnace by pyrophyllite mounts E. A thermo junction, not shown in the Figure 4.86, in contact with A serves to measure the temperature of the specimen. Temperatures of the order of 12000 K can be attained with a power input of the order of 10 W.

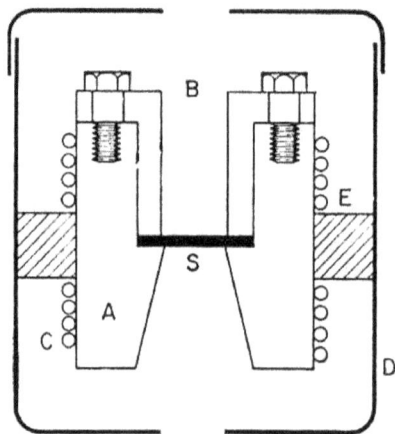

Figure 4.86   High and low temperature specimen mounts

The electron beam enters through the small aperture at the top of Figure 4.86. The larger aperture at the bottom gives ample clearance for the cone of diffracted rays from the specimen. A similar device can be used for cooling the specimen below room temperature. Instead of using a heating coil, a number of "frigistors" are attached to the specimen-mounting block. These are semiconductor junctions, which produce a large Peltier cooling when a current is passed through them.

To attain the lowest temperatures, the specimen is mounted in a block attached by a massive copper bar to a special double-walled vacuum vessel containing a liquid gas, the vacuum in the space between the walls being continuous with, and maintained by, the vacuum in the camera.

Photographic recording device   The fluorescent screen is used to view the diffraction patterns through the window W. To make a permanent record, photographic plate P is used. Simple arrangement is to mount the photographic plate in a light-tight box, which can be inserted into a vacuum port at the bottom of the camera. The fluorescent screen is attached to the hinged lid of the box, and can be lifted from outside the camera by rotating a shaft, which passes through an O-ring seal. The pattern can thus be focused on the fluorescent screen, and then, by raising the latter, the photographic plate can be exposed. Figure 4.87 shows the column of the modern electron diffraction camera ER-100.

Electron diffraction pattern   In earlier days, direct methods are used to solve the small molecular crystal structures. An electron diffraction pattern obtained, by the method of a selected area in an electron microscope, from copper perchloro phtalocyanine is shown in Figure 4.88a and projection of the potential of this structure (Figure 4.88b). All the atoms are clearly seen and the phases $|\phi_{hkl}|$ and structure amplitudes are determined by direct methods.

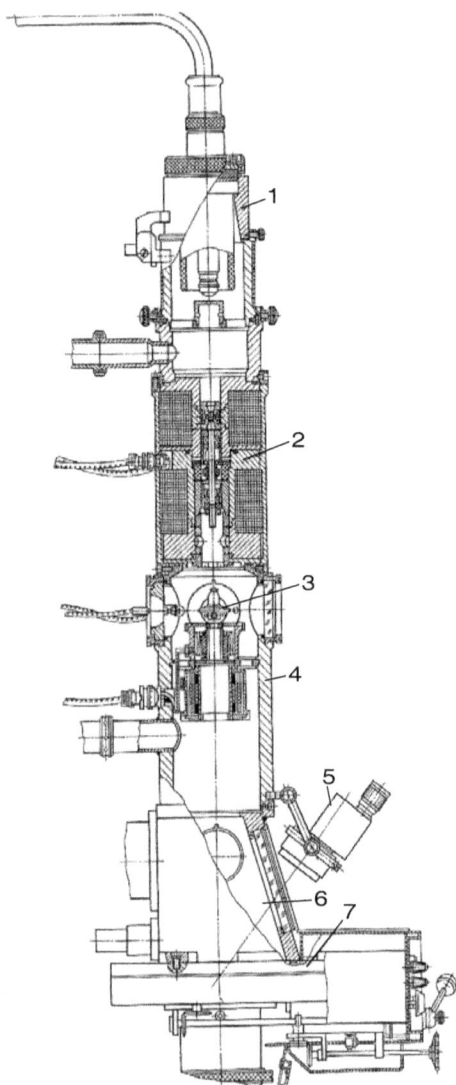

Figure 4.87 Modern electron diffraction camera ER-100 (1) electron gun
(2) electromagnetic lenses (3) specimen stage (4) chamber
(5) optical microscope for observing the image on the screen
(6) tube (7) photographic camera

The structures of many layer ionic lattices, crystal hydrates and hydroxides, organic and inorganic compounds containing hydrogen atoms have been studied by the electron diffraction method. It is widely used in analysing layer silicates, including clay minerals.

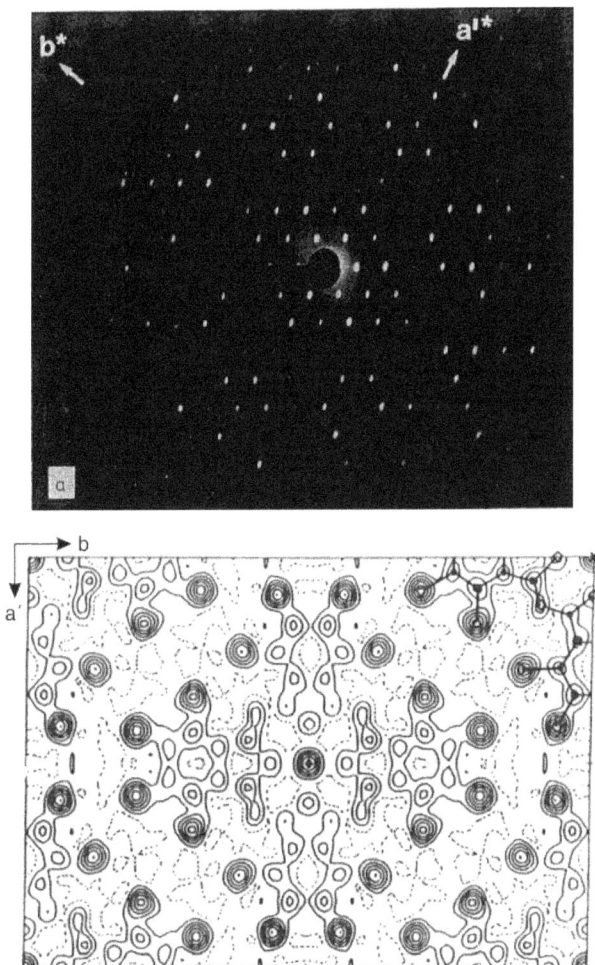

Figure 4.88 Study of copper perchlorophthalocyanine; (a) Electron diffraction pattern (b) Electrostatic potential map

Electrostatic potential map   The electron scattering is less dependent on the atomic number Z than X-ray scattering. This facilitates to distinguish, in the Fourier synthesis, potentials of light atoms in the presence of the heavy ones, including hydrogen atoms. Figure 4.89 shows the Fourier synthesis of the potential of the structure of paraffin in which we can see the position of hydrogen peaks clearly.

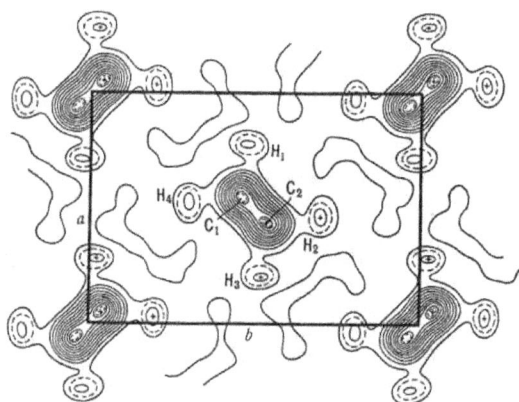

Figure 4.89 Fourier synthesis of the potential of structure of paraffin. Hydrogen atoms are clearly seen

Electron diffraction differs from X-ray and neutron scattering from its ability to form images and small probes. Equally important, electrons interact with matter through the strong coulomb potential. The first electron crystallographic protein structure to achieve atomic resolution was **bacteriorhodopsin** determined by Richard Henderson and coworkers at the Medical Research Council Laboratory of Molecular Biology in 1990. Since then, several other high-resolution structures have been determined by electron crystallography, including the light-harvesting complex, the nicotinic acetylcholine receptor, and the bacterial flagellum. About 143 protein structures were solved by cryo-electron microscopy and are deposited in protein data bank till March 2007.

NEUTRON DIFFRACTION

Neutron diffraction is a crystallographic method for the determination of the atomic structure of a material. The technique is similar to X-ray diffraction but the different type of radiation gives complementary information. Neutrons interact with matter differently than X-rays. X-rays interact primarily with the electron cloud surrounding each atom. On the other hand, neutrons interact directly with the nucleus of the atom, and the contribution to the diffracted intensity is different for each isotope; for example, regular hydrogen and deuterium contribute differently. It is also often the case that light (low Z) atoms contribute strongly to the diffracted intensity even in the presence of large Z atoms. *In the case of X-ray diffraction, hydrogen, having a single electron, scatters very weakly and is swamped by the heavier atoms, C, N, and O.* However, hydrogen atoms play important roles in biomolecules (in the formation of hydrogen bonds) and knowledge of their positions is desirable. The observation of hydrogen atoms is possible in case of neutron diffraction.

Neutron diffraction has become a useful tool because intense neutron beams are available from reactors. These neutrons are in thermal equilibrium with the moderator and their spectrum is approximately Maxwellian, with the temperature of the moderator. At 300 K, the mean energy is about 0.025 eV and the wavelength is 1.8 Å.

Neutrons interact with the nucleus via the nuclear force and with the magnetic moment of unpaired electrons via the electromagnetic one. The magnetic interaction permits the investigation of magnetic materials. Neutron diffracting investigations require powerful neutron sources. This purpose is served by high flux, slow-neutron nuclear reactors; pulsed reactors can also be used. According to de Broglie's equation, the wavelength is

$$\lambda = \frac{h}{mv} = \frac{h}{\sqrt{3mkT}}$$

where,

m is the mass of neutrons,

v is its velocity,

h and *k* are Planck's and Boltzmann's constant respectively, and

T is the absolute temperature.

The spectrum of the neutron beam channelled from the reactor is continuous ("white"), because of the Maxwellian velocity distribution; its maximum at 100°C corresponds to $\lambda = 1.3$ Å.

When it is necessary to use long-wave neutrons, the entire spectrum can be shifted with respect to energies by passing the reactor neutrons through cooling filters. These may take the form of chambers filled with liquid helium, hydrogen, or another moderator (e.g. beryllium) cooled to helium temperatures.

Neutron Diffractometer

In modern research reactors, a thermal-neutron flux of about 10^{15} cm^{-2}s^{-1} is maintained in the core. But the collimated flux of monochromatic neutrons that hits the specimen has a substantially lower intensity. Figure 4.90 is a schematic representation of a neutron diffraction unit. A neutron beam with a "white" spectrum passes through the reactor shielding along a channel terminating in a monochromator. Primary collimation is performed in the channel. Large single crystals of Cu, Zn, Pb, or other metals, or pyrolytic graphite plates usually serve as monochromators. The intensity of the resulting monochromatic beam strongly depends on the quality of the monochromator and also on the collimation; the monochromatic neutron flux is $10^7–10^8$ cm^{-2}s^{-1}. New possibilities for condensed matter research are opened by the use of powerful sources of pulsed neutrons. In the spallation process, neutrons are generated in the bombardment of a heavy-element (e.g. uranium) target by protons accelerated in a proton synchrotron. Such sources provide high-resolution neutron diffraction patterns and allow the conduction of various investigations by neutron techniques.

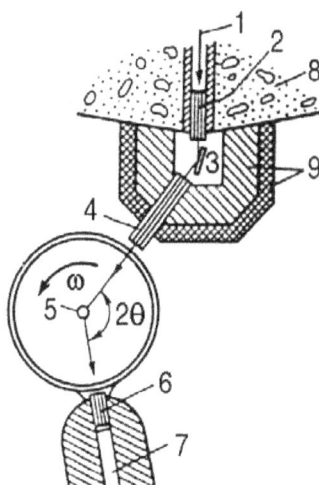

Figure 4.90 Neutron diffractometer 1. neutron beam from a reactor 2. primary collimator 3. monochromator 4. secondary collimator 5. specimen 6. collimator in front of the detector 7. neutron detector

Diffractometric units are similar in their principle and design to X-ray devices, but they are usually larger because the detector must have a thick radiation shielding. Proportional gas counters filled with ^{3}He or $^{10}BF_3$ are generally used as detectors. For polycrystalline specimens, it is sufficient to have a one-circle diffractometer, while for single crystals the four-circle design is most convenient. The devices are either fully automated or remote controlled. When required, various attachments are used; for cooling and heating of specimens, their magnetization, uniform compression, etc. Since the initial flux and the neutron scattering cross sections are less than for X-rays, the objects under investigation are larger than in X-ray investigations—several millimeters. In the polychromatic version, neutron diffraction can be used by analogy with the Laue X-ray method. Then, with the detector fixed and the crystal rotating, the reflected neutron beams with different λ can be measured by the time-of-flight method.

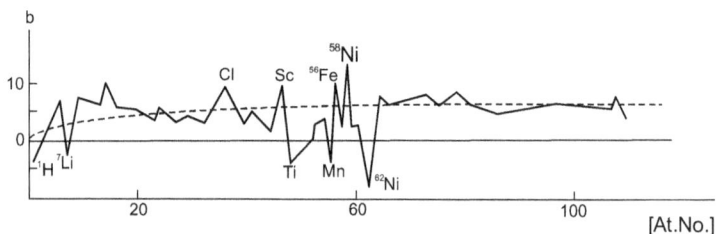

Figure 4.91   Dependence of the amplitude of coherent nuclear scattering of neutrons (—) on the atomic weight of the elements (----) scattering on the nuclear potential

Nuclear interaction is described by the amplitudes of nuclear scattering b, which are of the order of 10^{-12} cm and are measured in Fermi units f ($f = 10^{-13}$cm). The values of b vary non-monotonically with the atomic number Z (Figure 4.91). Isotopes of one and the same element have different values of b; for some isotopes, the value of b is negative because of the presence of resonance levels in their nuclei (this is not the case in either X-ray or electron scattering). Thus, for hydrogen ^1H_b = −3.74, for deuterium ^2D_b = 6.57 for carbon $^{12}C_b$ = +6.6, for nitrogen $^{14}N_b$ = +9.4, and for manganese $^{55}Mn_b$ = –3.7 *f*. Since the size of a nucleus is small (10^{-13} cm) as compared with the wavelength of the neutron λ = 10^{-8} cm, the values of *b* do not decrease with increasing scattering angle, i.e., they are constant for all sin θ/λ. Atoms or ions which have a non-zero spin and/ or orbital magnetic moment exhibit an additional interaction with the magnetic moment of the neutron, which is of the same order of magnitude as the nuclear interaction. The atomic amplitude f_M of magnetic scattering depends on the shape of the relevant electron shell and decrease with increasing sin θ/λ. The temperature factor is taken into account similar to X-rays. Besides, the effects of absorption, the inelastic coherent and incoherent scattering take place. The intensity of coherent elastic scattering of non-polarized neutrons by a crystal is determined by the sum of the squares of the structure amplitudes.

INVESTIGATION OF THE ATOMIC STRUCTURE

Neutron diffraction analysis is used predominantly for refining or obtaining additional information on structures studied by the X-ray method. Investigation is often conducted simultaneously with X-ray studies, and thus data on the unit cell, symmetry, and positions of most of the atoms are already available. Then the calculation of phases permits construction of the Fourier synthesis of nuclear density

$$n(r) = \sum_H F_{nH} \exp[-2\pi i(r.H)] \qquad (4.87)$$

In the absence of magnetic scattering, the peaks of this synthesis give time–average distribution of the nuclei due to the thermal motion; the peak heights are proportional to the scattering amplitudes *b* of the corresponding nuclei and, if *b* is negative, the peak is negative as well, i.e., it shows the atom as a "pit" on the Fourier synthesis (Figure 4.92a).

The position of the nuclei can be refined using difference synthesis and the least squares method with an isotropic temperature factors. Neutron diffraction data are especially suitable for the latter method because, as indicated above, the values of b are constant and the intensity decrease is solely due to the thermal motion.

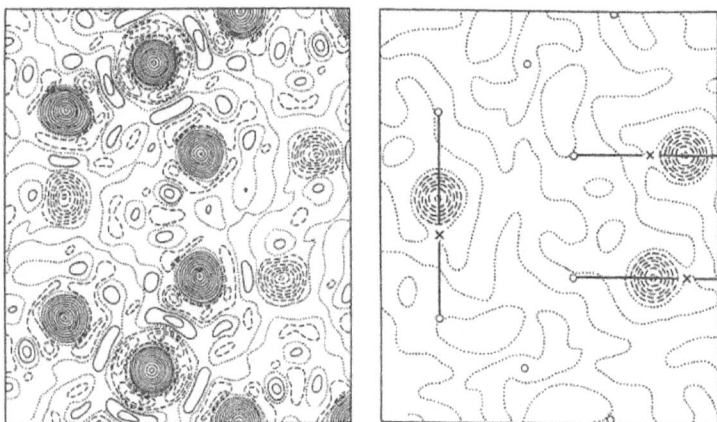

Figure 4.92 Structure of KH_2PO_4 in the ferroelectric state at $-180°C$. (a) Fourier projection of the nuclear density onto the (001) plane (b) projection of the difference synthesis on which H atoms are clearly visible (when the sign of the external electric field is reversed, the H atoms shift to the positions marked with crosses)

The advantages offered by neutron diffraction structure analysis are due to previously described features of nuclear amplitudes b. These include, in the first place, a better (as compared with X-ray analysis) possibility for determining the position of the light atoms in the presence of heavy ones. A significant advantage is the detection of hydrogen atoms in crystals of different compounds. Hydrogen atoms can be replaced, completely or partly, by deuterium, which provides additional information. The H peaks in the Fourier synthesis maps are negative and those of D are positive, in accordance with the sign of the amplitudes of scattering (Figure 4.93a).

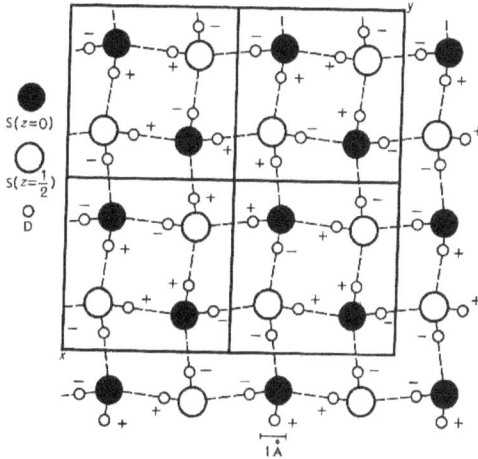

(a)

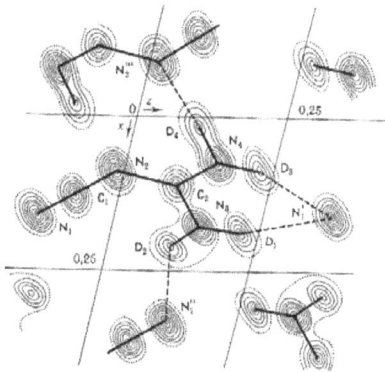

(b)

Figure 4.93 (a) Structure of solid D_2S at 102 K. Projection onto the (001) plane (---) hydrogen bonds forming zigzag chains parellel to the [100] and [010] axes (b) Projection of the nuclear density of the crystalline structure of deuterated dicyandiamide $C_2N_4D_4$. Dashed lines join the hydrogen-bonded atoms.

Various modification of ordinary and heavy ice, a number of crystal hydrates, many organic and inorganic compounds including metal hydrides, hydrogen containing ferroelectrics, and phase transitions in them were studied in this way. Other examples of investigated structures with atoms differing drastically in their atomic numbers Z include nitrides, carbides, oxides of heavy metals, etc. The structure of solid D_2S at 102 K is shown in Figure 4.93a and the nuclear density of deuterated dicyandiamide $C_2N_4D_4$ is shown in Figure 4.93b.

Another advantage of neutron-diffraction studies consists of the possibility of investigating structures containing atoms with close Z which are almost indistinguishable by the X-ray method. Examples are Fe, Ni, Co and Cr alloys and their compounds, for instance, ferrospinels, complicated oxides, and silicates containing Mg and Al. The amplitudes b for such atoms or their isotopes differs widely enough for the individual positions of these atoms to be determined. The difference in b for isotopes of a given element makes it possible, in principle, to investigate the ordering of isotopic nuclei in crystalline structures.

Since the values of b are independent on the scattering angle, the decrease in structure amplitudes F_{nH} with increasing $|H|$ depends exclusively on the temperature factor. Therefore the neutron structure amplitudes can be measured up to higher values of $\sin \theta/\lambda$, i.e., higher hkl (lower d_{hkl}) than in X-ray or can give both the positional and thermal motion parameters with an accuracy higher than in X-ray diffraction. This is used, in particular, when constructing difference X-ray neutron diffraction synthesis.

Neutron diffraction studies open up additional opportunities for determining crystal structures involving comparison of F_x, F_n, the structure amplitude in case of X-ray and neutron diffraction, respectively. As the X-ray and neutron atomic (nuclear) amplitudes of individual atoms appearing in the structure–amplitude equation are different, such a comparison is equivalent to isomorphous replacement. In addition, for neutron there is an effect similar to anomalous X-ray scattering; thus we can determine the position of the anomalously scattering atoms using neutron diffraction.

In some cases the "zero" matrix method, i.e., the use of isotopes or different atoms with opposite signs of b, may prove to be highly sensitive. If their nuclei occupy equivalent positions of the unit cell, such positions can be "left out" of diffraction by an appropriate choice of concentrations. Thus, only diffraction from other atoms takes place.

The most complicated compounds investigated by neutron diffraction are vitamin B_{12} and myoglobin.

REVIEW QUESTIONS

1. Explain the term reciprocal lattice.

2. What is the significance of Ewald sphere?

3. How are X-rays produced?

4. What are characteristic X-rays?

5. Why are X-rays used for structure determination of compounds?

6. Explain Laue's equations.

7. Explain Bragg's law and derive the condition for diffraction maximum.

8. Explain rotating crystal method.

9. Explain the principle used in powder diffraction.

10. What are the advantages and disadvantages of powder diffraction methods?

11. Define scattering factor. Explain the scattering curve for copper.

12. What is structure factor?

13. How do we get structure factor amplitudes from the measured intensities of Bragg's reflections?

14. What is Fourier transform?

15. What is phase problem in crystallography?

16. Explain the various methods to solve the phase problem.

17. Differentiate the electron density map and Patterson map.

18. Explain the principle used in isomorphous technique.

19. What is Friedel's law?

20. What is anomalous dispersion?

21. Explain the triplet and quartet relations in direct methods.

22. How is least-squares refinement procedure used in crystal structure analysis?

23. Define R-factor.

24. What do you mean by $\pi...\pi$ and C–H$...\pi$ interactions?

25. What are the possible conformations of a five-membered ring?

26. Write the differences between X-ray diffraction and neutron diffraction techniques.

27. Explain the basic principle in electron diffraction.

28. What are the advantages of neutron diffraction?

Problem 1 Show that the reciprocal lattice to fcc lattice is a bcc lattice. [Given the primitive translation vectors $a' = 1/2a(\hat{x} + \hat{y}); \ b' = 1/2a(\hat{y} + \hat{z}); \ c' = 1/2a(\hat{z} + \hat{x})]$.

Solution

$$a' = \frac{1}{2}a(\hat{x} + \hat{y})$$

$$b' = \frac{1}{2}a(\hat{y} + \hat{z})$$

$$c' = \frac{1}{2}a(\hat{z} + \hat{x})$$

∴ Volume of the fcc unit cell $V = |a'.b' \times c'|$

$$= \frac{1}{2}\left| \frac{1}{2}a(\hat{x} + \hat{y}) \cdot \frac{1}{2}a(\hat{y} + \hat{z}) \times \frac{1}{2}a(\hat{z} + \hat{x}) \right|$$

$$= \frac{a^3}{4}$$

$$a^* = 2\pi \frac{b' \times c'}{a' \cdot b' \times c'}$$

$$= 2\pi \frac{(a^2/4(\hat{x} + \hat{y} - \hat{z}))}{a^3/4}$$

$$= \frac{2\pi}{a}(\hat{x} + \hat{y} - \hat{z})$$

$$b^* = 2\pi \frac{c' \times a'}{a' \cdot b' \times c'}$$

$$= 2\pi \frac{(a^2/4(-\hat{x} + \hat{y} + \hat{z}))}{a^3/4}$$

$$= \frac{2\pi}{a}(-\hat{x} + \hat{y} + \hat{z})$$

$$c^* = 2\pi \frac{a' \times b'}{a' \cdot b' \times c'}$$

$$= 2\pi \frac{(a^2 / 4(\hat{x} - \hat{y} + \hat{z}))}{a^3/4}$$

$$= \frac{2\pi}{a}(\hat{x} - \hat{y} + \hat{z})$$

Thus $\dfrac{2\pi}{a}(\hat{x} + \hat{y} - \hat{z})$, $\dfrac{2\pi}{a}(-\hat{x} + \hat{y} + \hat{z})$ and $\dfrac{2\pi}{a}(\hat{x} - \hat{y} + \hat{z})$ correspond to a bcc lattice.

Problem 2   Calculate the wavelength of X-rays that give third order reflection at an angle of 6° for (111) reflection from a cubic lattice of edge length 5.63 Å.

Solution

Given $n = 3$, $\theta = 6°$, $d_{hkl} = d_{(111)}$, $a = 5.63$ Å.

Bragg's law states that $2d \sin\theta = n\lambda$

$$d = \frac{a}{\sqrt{h^2 + k^2 + l^2}}$$

$$\therefore \quad \frac{2a \sin\theta}{\sqrt{h^2 + k^2 + l^2}} = n\lambda$$

$$\lambda = \frac{2a \sin\theta}{n\sqrt{h^2 + k^2 + l^2}}$$

$$= \frac{2 \times 5.63 \times 10^{-10} \times \sin 6°}{3 \times \sqrt{1 + 1 + 1}}$$

$$= 0.2265 \text{ Å}$$

The wavelength of X-rays that give third order reflection at an angle of 6° is found to be 0.2265 Å.

Problem 3 The powder pattern of a cubic crystal show the reflections at the following angles: 5°39', 8° and 9°49'. Identify the lattice type.

Solution

Bragg's reflection angles: 5°39', 8° and 9°49'.

From Bragg's law of reflection

$$2d \sin \theta = n\lambda$$

$$d_{hkl} = \frac{n\lambda}{2 \sin \theta}$$

$$\therefore d_{100} : d_{110} : d_{111} = \frac{n\lambda}{2 \sin \theta_1} : \frac{n\lambda}{2 \sin \theta_2} : \frac{n\lambda}{2 \sin \theta_3}$$

$$= \frac{1}{\sin \theta_1} : \frac{1}{\sin \theta_2} : \frac{1}{\sin \theta_3}$$

$$= \frac{1}{\sin 5°39'} : \frac{1}{\sin 8°} : \frac{1}{\sin 9°49'}$$

$$= 10.1573 : 7.1853 : 5.8652$$

$$= 1 : 0.074 : 0.5774$$

This ratio corresponds to sc type of crystal lattice. Hence the given pattern is found to be from sc type of lattice.

Problem 4 The wavelength of Cu K_α X-rays is 1.54 Å. What is the minimum voltage that must be applied between cathode and anode (the target) of an X-ray tube to produce these X-rays?

Solution

The energy of an electron (charge, e) accelerated through a voltage drop V is eV. If we neglect small energy losses in the target, the X-ray photon will have energy

$$h\upsilon = \frac{hc}{\lambda} = eV$$

$$V = \frac{hc}{\lambda e}$$

$$= \frac{(6.63 \times 10^{-34} \, JS)(3 \times 10^{8} \, ms^{-1})}{(1.54 \times 10^{-10} \, m)(1.6 \times 10^{-19} \, c)}$$

$$= 8.072 \, kV$$

Thus the minimum voltage required to produce X-rays of wavelength 1.54 Å is 8 kV.

Problem 5   Derive the condition for the allowed Miller indices for a face-centred crystal (say NaCl).

Solution

In NaCl

Na^+ ions are at 0, 0, 0; 1/2, 1/2, 0; 1/2, 0, 1/2 and 0, 1/2, 1/2

Cl^- ions are at 1/2, 0, 0; 0, 1/2, 0; 0, 0, 1/2 and 1/2, 1/2, 1/2

$$F_{hkl} = f_{Na^+} \left\{ \exp[2\pi i(0)] + \exp\left[\frac{2\pi i(h+k)}{2}\right] + \exp\left[\frac{2\pi i(h+l)}{2}\right] + \exp\left[\frac{2\pi i(k+l)}{2}\right] \right\} +$$

$$f_{Cl^-} \left\{ \exp\left[\frac{2\pi ih}{2}\right] + \exp\left[\frac{2\pi ik}{2}\right] + \exp\left[\frac{2\pi il}{2}\right] + \exp\left[\frac{2\pi i(h+k+l)}{2}\right] \right\}$$

$$\exp\left(\frac{2\pi in}{2}\right) = \cos \pi n + i \sin \pi n$$

If n is any integer

$$\cos \pi n = (-1)^n$$

$$\therefore \exp\left(\frac{2\pi in}{2}\right) = (-1)^n$$

$$\therefore \text{ If n is an even integer, } \exp\left(\frac{2\pi i n}{2}\right) = +1$$

$$\text{If n is an odd integer, } \exp\left(\frac{2\pi i n}{2}\right) = -1$$

Thus $F_{bkl} = f_{Na^-}\{1 + (-1)^{b+k} + (-1)^{k+l} + (-1)^{b+l}\}$

$$f_{cl^-}\{(-1)^b + (-1)^k + (-1)^l + (-1)^{b+k+l}\}$$

The term $[1 + (-1)^{b+k} + (-1)^{k+l} + (-1)^{b+l}]$ is zero unless the indices are either all odd or all even.

Thus fcc type of crystal lattice is detected by the systematic absences of reflections of mixed type such as (210) or (110). Thus condition for allowed indices are b, k, l should be either all odd or all even.

Problem 6 Deduce the systematic absence conditions for a two-fold screw axis parellel to b axis of a unit cell.

Solution

The equivalent positions related by the screw axis are x, y, z and $\overline{x}$, $y + 1/2$, $\overline{z}$

$$\therefore F_{bkl} = fe^{2\pi i(b\overline{x}_j + k\overline{y}_j + l\overline{z}_j)}$$

$$F_{0k0} = f[e^{2\pi i(0(x) + k(y) + 0(z))}] + f[e^{2\pi i(0(\overline{x}) + k(y+1/2) + 0(\overline{z}))}]$$

$$= f[e^{2\pi i ky} + e^{2\pi i k(y+1/2)}]$$

$$F_{bkl} = fe^{2\pi i ky}\left[1 + e^{\frac{2\pi i k}{2}}\right]$$

$$= fe^{2\pi i ky}[1 + (-1)^k]$$

Thus $(0k0)$ reflections are absent when k is an odd integer which is the systematic absence condition for given two-fold screw-axis.

EXERCISES

1. A unit cell has the dimensions $a = 4$ Å, $b = 6$ Å, $c = 8$ Å, $\alpha = \beta = 90°$, $\gamma = 120°$. Determine a) a^*, b^*, c^* for the reciprocal cell and b) volume of the real unit cell.

   (Ans: 0.29, 0.17, 0.14 Å, 166 Å³)

2. Electrons are accelerated by 844 volts and the generated X-rays are reflected from a crystal. The reflection maximum occurs when the glancing angle is 58°. Determine the spacing of the crystal.

   (Ans: 0.0248 nm)

3. The $\sin^2 \theta$ values observed for a sample MgO powder with 0.71 Å X-rays are: 0.02134, 0.02857, 0.05734, 0.07841, 0.08613, 0.11437, 0.13671 and 0.14358. To which type of cubic lattice do the data correspond? Calculate the edgelength of the unit cell. The density of MgO is 3.58 g cm⁻³.

   (Ans: fcc type; $a = 4.21$ Å)

4. Derive the condition for the allowed Miller indices for a body-centred crystal.

   (Ans: $h + k + l$ = even absent)

5. Deduce the systematic absence conditions for a glide plane type 'c' reflecting in a b axis.

   (Ans: ($h0l$) odd absent)

GLOSSARY

Anomalous scattering When the incident wavelength is close to the absorption edge of an atom, the scattering factor beomes a complex quantity $f = f_0 + f' + i f''$. This phenomenon is known as anomalous scattering.

Atomic packing factor Sometimes called a packing fraction, it is the fraction of volume in a crystal structure that is occupied by atoms. It is dimensionless and always less than unity.

Atomic scattering factor The ratio of the amplitude of the wave scattered by an atom to the amplitude of the wave scattered by an electron. It is also called atomic form factor.

Basis A group of atoms, ions or molecules around a lattice point.

Bond energy The amount of energy necessary to break one mole of covalent bonds into isolated gaseous species; it is sometimes called the bond dissociation energy or bond enthalpy.

Bond length The distance between the centres of the nuclei involved in the bond.

Bond order A measure of how many pairs of electrons are shared between two atoms to bond those two atoms together.

Bragg's law The Law states that the condition for diffraction maximum occurs only at a particular angle of incidence and also when the path difference between the two scattered waves is equal to integral multiples of the incident wavelength ($2d \sin \theta = n\lambda$).

Bravais lattice The lattice that consists of all lattice points as equivalent and all of same kind.

Calculated Structure factor The quantity F_c obtained by calculating A_{hkl} and B_{hkl} values for a model (using the known atomic positions) and hence $|F_{hkl}| = \sqrt{A_{hkl}^2 + B_{hkl}^2}$.

CCD detector Charge-coupled-device-based detector where frame of intensities are recorded.

Centrosymmetry The symmetry obtained when each point in the object is converted to an identical point by projecting through a common centre and extending an equal distance beyond this centre.

Coherent scattering The type of scattering in which the incident and scattered radiations have the same wavelength and have electric and magnetic fields that are "in phase" with the incident radiation.

Combined figure of merit An indicator to select the correct set from the multiple solutions.

Coulombic force The force between the two oppositely charged particles which is directly proportional to the product of their charges and inversely proportional to the square of the distance between them.

Covalent bond The bonds in compounds that result from the sharing of one or more pairs of electrons.

Crystal lattice A regular periodic arrangement of atoms, ions or molecules in three-dimensional space.

Delocalized electrons The electrons which move freely within the molecular orbitals and so each electron becomes detached from its parent atom.

Diamond glide Diamond glide occurs only in space groups with face centred or body centred cells, and is characterized by a translation of $(a + b)/4$, $(b + c)/4$, or $(c + a)/4$ after rotation.

Dipole Two equal and opposite charges separated by a smaller distance.

Direct methods Mathematical or probabilistic procedures using which phase values can be obtained from phase relationships directly from measured intensities.

Electron affinity The work done to add electron to the shell of an electronegative atom.

Electronegativity The relative ability of an atom to draw electrons in a bond towards itself.

Extinction correction This is a correction factor arising due to slightly misoriented crystal blocks.

Fourier synthesis A mathematical process to transform the diffraction pattern produced by the X-ray scattering of atoms into an image.

Friedel's law $I_{hkl} = I_{\overline{hkl}}$.

Glide planes Glide planes (reflection + translation) occur when a mirror operation is followed by a translation of a fraction of the unit cell parallel with the mirror plane.

Goniometer A part of X-ray diffractometer which allows the specimen to be precisely oriented in any position while remaining in the X-ray beam. It is also used in oscillation, rotation and precession photograph method.

Goodness of fit A statistical parameter whose value during the refinement procedure approaches 1 in the final stage.

H e r m a n n – M a u g u i n symbol A notation commonly used by crystallographers to represent the point groups (H–M symbol).

Hydrogen bond The strong electrostatic attraction that occurs between molecules in which hydrogen is in a covalent bond with a highly electronegative element.

Image plate These can be used as a replacement of the area detector. Image plates are exposed to X-rays, as to any other detector, and the X-ray photon causes a chemical change in the plate coating that releases a fluorescence that is detected by a photomultiplier when scanned with light of proper wavelength.

Interference The superposition (overlapping) of two or more waves resulting in a new wave pattern.

Intermolecular attraction Attraction between one molecule and a neighbouring molecule.

Intramolecular attractions The forces of attraction which hold an individual molecule together (for example, the covalent bonds) are known as intramolecular attractions.

Ionization potential Work done to remove an electron from the shell of an electropositive atom.

Isomorphous replacement A technique used in protein crystallography to determine the macro-molecular structure. Intensity is collected first for the crystal of the macromolecule (native data). The crystal is soaked in heavy-atom salt solutions (U, Gd, Se, Pt, Au, etc.) and derivative is formed. X-ray data is collected for the derivative. For the same Bragg reflections, isomorphous differences are observed by native and derivative data and are made use in determining the heavy atom positions.

Lattice A regular, infinite arrangement of points in which every point has the same environment as any other point is known as lattice.

Lorentz-correction Lorentz factor (L) has the simple form of

$1/\sin 2\theta$ expressing the fact that, for the constant angular velocity of the rotation of the crystal, different reciprocal lattice points pass through the sphere of reflection at different rates and thus have different time- of-reflection opportunity.

Madelung   constant A dimensionless constant which determines the electrostatic energy of a three-dimensional periodic crystal lattice consisting of a large number of positive and negative point charges when the number and magnitude of the charges and the nearest-neighbour distance between them is specified.

Miller indices   A set of numbers which quantify the intercepts of a crystal plane and which are used to uniquely identify the plane or surface.

Multiple-wavelength anomalous dispersion A technique used to determine a macromolecular structure, in which data collection at multiple wavelengths (three) is carried out using synchrotron radiation.

Non-Bravais lattice The combination of two or more interpenetrating Bravais lattices with fixed orientation relative to each other which contains non-equivalent lattice points.

Non-polar covalent bond A bond in which the electrons are shared equally between two atoms.

Non-primitive cell   The unit cell that contains more than one lattice point.

Normalized structure factor Normalized structure factor corresponds to structure factor of a point atom model.

Observed structure factor The quantity F_o obtained from the experimental measurement of Bragg intensities by the relation $|F_o| \alpha \sqrt{I_{hkl}}$.

ORTEP   Oakridge Thermal Ellipsoidal Plot, is obtained by the programme ORTEP which shows the three-dimensional picture of a molecule depicting the thermal vibration of the atoms in the molecule.

Oscillation   method The method of data collection in which a single large crystal is oscillated through a small angle about a fixed axis set normal to the X-ray beam and the diffraction pattern is recorded on a cylindrical camera in which a curved photographic plate is also placed normal to the X-ray beam.

Patterson synthesis   A synthesis used to locate the heavy atom. A patterson denotes interatomic vector only and is not an atomic site. The strength of the patterson is the product of the atomic numbers of the connected atoms. A unit cell made up of N atoms will have (N^2-1) Patterson peaks

$$P(u,v,w) = \sum_{hkl} I_{hkl} \cos 2\pi(hu + kv + lw).$$

Pauli's Exclusion principle Pauli Exclusion principle states that no two electrons can exist in the same quantum state.

Phase ambiguity When anomalous scattering is used to overcome the phase problem, the total phase $\alpha_N = \alpha_P + \theta + \dfrac{\pi}{2}$. θ can be calculated from the anomalous differences. The resultant two values of θ ($\pm\theta$) is known as phase ambiguity.

Phase problem The non-availability of phases of Bragg reflections, which are required along with their known structure factor magnitudes, to locate the electron density maxima (to locate eventually an atomic site) is known as phase problem in crystallography.

Point group A collection of symmetry operations that define the symmetry about a point (such that the symmetry operation leaves at least one point invariant).

Polar compounds Compounds for which the electronegativity difference is between about 1.2 and 1.8.

Polar covalent bond A bond in which one atom has a greater attraction for the electrons than the other atom.

Polarization correction The radiation from a normal X-ray tube is unpolarized, but after reflection from a crystal, the beam is polarized and the fraction of energy lost in this process is dependent only on the Bragg's angle. The polarization factor (p) is given by $p = \dfrac{1 + \cos^2 2\theta}{2}$. L and p factor correction is essential in order to bring the $|F|^2$ data on to a correct relative scale.

Polycrystalline Those samples that are made up of a number of smaller crystals known as crystallites.

Precession method Experimental method of calculating the reciprocal lattice parameters.

Primitive cell The unit cell that contains just only one lattice point.

Quartet relationship A four-phase relationship of the form $\phi_{\vec{h}}' + \phi_{\vec{k}}' + \phi_{\vec{l}}' + \phi_{\vec{m}}' = \phi_{\vec{h}} + \phi_{\vec{k}} + \phi_{\vec{l}} + \phi_{\vec{m}}$ such that $\vec{h} + \vec{k} + \vec{l} + \vec{m} = 0$.

Reciprocal lattice A set of imaginary points constructed in such a way that the direction of a vector from one point to another coincides with the direction of a normal to the real space planes and the separation of those points (absolute value of the vector) is equal to the reciprocal of the real interplanar distance.

R-factor   Residual factor involves adding together the absolute values of all the discrepancies between observed and calculated structure factor amplitudes, and normalizing the sum by dividing it by the sum of all the observed amplitudes, to give a value called reliability index of the proposed model (3% to 8%).

Schoenflies notation (S)   A notation commonly used by spectroscopists to represent the point groups.

Scintillation counter   A solid-state detector that rely on the fact that when an X-ray is incident onto certain materials, such as a crystal of sodium iodide activated with thallium, visible light is emitted which may be caught by a photomultiplier.

Screw axis   A screw axis (rotation + translation) occurs when a proper rotation axis operation is followed by a translation by a fraction of the unit cell in the direction of the rotation.

Secondary structure   The regular, repeated configurations caused by hydrogen bonding between atoms of the polypeptide backbone of proteins.

Single crystal   Also called monocrystal, it is a crystalline solid in which the crystal lattice of the entire sample is continuous and unbroken to the edges of the sample, with no grain boundaries.

Slow evaporation   A commonly used crystallization technique in which the sample is dissolved in a solvent and a saturated solution is prepared first and then subjected to evaporation.

Space group   The combination of point groups with translation symmetry performed on an object to represent the whole symmetry.

Structure factor   The contribution to the scattering by the unit cell contents (different atoms each having its own scattering factor f_j position at (x_j, y_j, z_j) the fractional coordinates of the jth atom).

Symmetry   The invariance of an object under some kind of transformation.

Symmetry element   A geometrical entity such as a point, line, or plane about which a symmetry operation is performed.

Torsion   angle   When considering four atoms connected in the order A–X–Y–B, the rotation about the X–Y bond is known as torsion angle. It is also equal to the angle between the two normals to the planes formed by A–X–Y and X–Y–B.

Triplet relation   A three-phase relationship of the form
$$\phi_{\vec{h}}' + \phi_{\vec{k}}' + \phi_{\vec{l}}' = \phi_{-\vec{h}} + \phi_{-\vec{k}} + \phi_{-\vec{l}}$$
such that $\vec{h} + \vec{k} + \vec{l} = 0$.

Violation of Friedel's law   In the presence of anomalous scattering, $I_{hkl} \neq I_{\overline{hkl}}$.

Wilson plot   The graphical method used to obtain the overall scaling factor and temperature factor.

REFERENCES

Ali Omar, M. (2002). *Elementary Solid State Physics: Principles and Applications.* Pearson Education, (Singapore) Pvt. Ltd.

Azaroff, L.V. (1960). *Introduction to Solids.* McGraw-Hill, New York.

Bacon, G.E. (1962). *Neutron Diffraction.* Oxford University Press, Oxford.

Buerger, J.M. (1960). *Crystal Structure Analysis.* John Wiley, New York.

Buerger, M.J. (1964). *The Precession Method in X-ray Crystallography.* Wiley, New York.

Cullity, B.D. (1972). *Elements of X-Ray Diffraction.* Freeman; Addison-Wesley. New York.

Dennis Sherwood (1976). *Crystal, X-rays and Proteins.* Longman Group Limited, London.

Donald, E., Sands. (1969). *Introduction to Crystallography.* Benjamin, W.A. Inc. New York.

Glusker, J.P. and Trueblood, K.N. (1985). *Crystal Structure Analysis: A Primer,* 2nd edn. Oxford University Press.

Giacovazzo, C., Monaco, H.L., Viterbo, D., Scordari, F., Gulli, G., Zanotti, G. and Catti, M. (1992). *Fundamentals of Crystallography.* Oxford University Press.

Kittel, C. (1977). *Introduction to Solid State Physics,* 5th edn. Wiley Eastern Limited.

Ladd, M.F.C. and Palmer, R.A. (1993). *Structure Determination by X-ray Crystallography*, 3rd edn. Plenum, New York.

Martin, J. Buerger (1967). *Crystal Structure Analysis*. John Wiley & Sons Inc. New York.

Mckie, D. and Mckie, C. (1986). *Essentials of Crystallography*. Blackwell, Oxford.

Parthasarathy, S. and Jenny, P. Glusker (1997). *Aspects of Crystallography in Molecular Biology*. New Age International Publishers.

Pillai, S.O. (1996). *Solid State Physics*. New Age International (P) Ltd. New Delhi.

Rymer, T.B. (1970). *Electron Diffraction*. Methuen & Co. Ltd., London.

Stout, G.H. and Jensen, L.M. (1968). *X-Ray Structure Determination: A Practical Guide*. Macmillan, New York.

James, R.W. (1967). *The Crystalline State* –Vol. II. G. Bell & Sons Ltd., London.

William Clegg (1998). *Crystal Structure Determination*. Oxford University Press.

Woolfson, M.M. (1997). *An Introduction to X-ray Crystallography*, 2nd edn. Cambridge University Press.

Dunitz, J.D. (1995). *X-ray Analysis and the Structure of Organic Molecules*. Wiley-VCM, Weinheim.

Karle, J. and Hauptman, H. (1956). "Structure invariants and semiinvariants for non-centrosymmetric space groups." *Acta Cryst.* 9: 45–55.

Karle, J. and Hauptman, H. (1956). "A theory of phase determination of the four types of non-centro symmetric space groups, 1p222, 2p22, 3p$_1$2, 3p$_2$2." *Acta Cryst.* 9: 635–651.

Hauptman, H. (1975). "A new method in the probabilistic theory of the structure invariants." *Acta Cryst.* A31: 680–687.

Lessinger, L. and Wondratschek, H. (1975). "Semiinvariants for space groups I2m and I2d." *Acta Cryst.* A31: p. 521.

Karle, J. and Hauptman, H. (1953). "The probability distribution of the magnitudes of a structure factor. I. The centrosymmetric crystal." *Acta Cryst.* 6: 131–135.

Hauptman, H. and Karle, J. (1953). "The probability distribution of the magnitudes of a structure factor. II. The non-centrosymmetric crystal." *Acta Cryst.* 6: 136–141.

Hauptman, H. and Green, E.A. (1976). "Conditional probability distributions of the four-phase structure invariant $\phi_{\vec{h}} + \phi_{\vec{k}} + \phi_{\vec{l}} + \phi_{\vec{m}}$ in PT." *Acta Cryst.* A32: 45–49.

Cochran, W. (1955). "Relations between the phases of structure factors." *Acta Cryst.* 8: 473–478.

Sheldrick, G.M. (1997). *SHELXS97* and *SHELXL97*. University of Gottingen, Germany.

Spek, A.L. (1990). "*PLATON*—A multipurpose crystallographic tool." *Acta Cryst.* A46, C34.

Spek, A.L. (2003). *J. Appl. Cryst.* 36: 7–13.

Hauptman, H. (1976). "The role of cosine semi-invariants." Plenum Press.

INDEX

www.ingramcontent.com/pod-product-compliance
Lightning Source LLC
Chambersburg PA
CBHW031806190326
41518CB00006B/214